The Greatest Safari

A portion of the proceeds from the sale of this book
will be donated to community work in Welverdiend village

First published by Gyldendal in 2011 as:
I begyndelsen var Afrika ... historier om livets udvikling set gennem natur og mennesker i Afrika
(978-8-702107-56-2)

This edition published in 2014 by:

SOUTHBOUND
an imprint of 30° South Publishers (Pty) Ltd.

16 Ivy Road
Pinetown
Durban 3610
South Africa

www.albatros-travel.com
sr@albatros-travel.dk
www.30degreessouth.co.za

English translation by Niel Standford & Søren Rasmusen
Designed & typeset by SA Publishing Services, South Africa (kerrincocks@gmail.com)
Cover design by Kerrin Cocks
Printed in South Africa by Pinetown Printers, Durban, KwaZulu-Natal

ISBN 978-1-928211-51-8
ebook 978-1-928211-55-6

The Greatest Safari

… in the beginning was Africa

The story of evolution seen from the
African savannah

Søren Rasmussen

Contents

Introduction   6
The Honeyguide's Tale   16
White Man in Africa   24
Safari in the Masai Mara   41
An Odyssey through the Evolution of Evolution   75
Philosophical Heavyweights   97
A Bridge between Spirit and Matter   120
Animals and Plants – Parallel Stories   144
A Detour to the Plant Kingdom   158
Nature Invents Social Networks   181
King Kong in the Mist:
   On the Track of Gorillas, Human Apes and Human Beings   211
Where Did We Come From?   242

Sources   255
Aknowledgements   255

TWO mutually quite unrelated events in 1984 turned into the beginning of a twenty-five-year odyssey which has resulted in this book.

The first event was in January, when, for the first time, I found myself in Kenya investigating the opportunities for safaris. The day after my arrival we roared off in Betsy, the old Land Rover, to the Masai Mara, where we pitched a primitive camp on a bluff by the Mara River under the characteristic Oloololo Escarpment. We were three people in two primitive boy-scout tents. Immediately after our arrival, the others decided to drive to Mara River Camp to repair the car, which had seen better days, while I had the pleasure of watching over our indispensable tents and their contents.

The rear lights of the Land Rover disappeared slowly into the darkness, while I sat in something resembling a Wegner chair by the blazing fire with a massively heavy Masai spear in my hand. Not long before, the two seasoned safari specialists had instructed me that in the event of a lion attack I should simply go down on my knee with the blunt end of the spear towards the ground, after which the charging monster would, for all the world, leap and impale itself on the spear point. The sounds were intense. A large pride of lions surrounded the camp, while leopards were slinking around in the trees, and the hyenas were waiting patiently in the background for the pitiful leftovers. I walked slowly backwards into the tent and zipped down. The tent was supposed to be absolutely secure, but the spear accompanied me inside.

There followed several eventful days,

but let me confine myself to saying I was sold. Sold to safari life, even though naturally, I did not know it was the prelude to a hundred safari journeys and 1,000 safari days over the years that followed, in which I arranged, sold and led safaris.

The second event took place immediately before the first dramatic safari experience: I completed my biology dissertation on viruses which infect bacteria. Since the first aquarium of my childhood, biology has been a part of my life, closely associated with Wellington boots, butterfly nets and binoculars, and at university I was briefly enamoured with the white laboratory coat, even though it was without a stethoscope.

To think that there also existed viruses which specialised in bacteria. It also turns out that a virus can choose between several different life strategies, depending on the environmental factors, even though they possess neither brain nor the basic life-functions. Inside a foreign cell they can either hasten to make a mass of copies of themselves and send them on – this is often what makes the patient ill – or they can also cut their way into the genes of the organism, into the chromosomes, and, finally become a part of the organism in question's genome which is passed on from generation to generation.

At university we were not of the opinion that viruses could be classified as living organisms, because they cannot independently maintain the essential life functions such as breathing and reproduction. They require the cells and genes of other organisms. And how would our life functions respond outside our regular environment? Even further: how much will two human beings be able to reproduce stuck on an ice floe in the Antarctic?

Microorganisms populated the Earth billions of years before us; they take up far more room than us and have adapted to every possible living space, deep down in the ground, under Antarctic ice floes, in spouting geysers and bubbling sulphur springs. We don't really know a great deal about them, and we should be wary of the notion that it is we who are in control.

These two events – the experience of Africa's wild nature and the discovery that the boundaries of life are fluid – constitute the thread that runs throughout the following pages. Before I withdraw into the background and let the story of life and the development of the human being roll over stock and stone, it should, for the sake of completeness, be mentioned that in the intervening 25 years I have reinforced my connection with Africa through the establishment of a number of safari agencies and savannah hotels, from which many stories and events are drawn. The significant pivotal point of the book is Kenya's famous nature reserve, the Masai Mara, and the modest safari lodge, Karen Blixen Camp, which was completed in 2007.

Since my first visit I have returned to the Masai Mara every single year, often with my family, in which our children have been obliged to learn the

blessings of camp-life from the age of six months. Conditions have improved in the course of time. Our first Karen Blixen Camp consisted of portable tents with 'longdrops' that replaced the spade and toilet rolls. There followed a so-called semi-permanent camp with most facilities provided, and, most recently the new Karen Blixen Camp which is, in some sense, the nearest one comes to luxury in the bush. Let me admit that I do enjoy it.

For more than twenty-fivr years I have had the privilege of being able to travel back and forth in time, as well as back and forth between Denmark and Africa, as I've said, more than 100 times, I have worn out several pairs of hiking boots on Kilimanjaro, Mount Kenya and the Drakensberg, driven kilometre after kilometre over the African savannah and spent thousands of hours in a safari chair with binoculars and a stack of books to hand. Books form the journeys' second component: excess weight on the trip out and double overweight on the journey home. Plants, mammals, acacias, birds of Africa, Næsbø, Hamsun, Joyce, Blixen and Kierkegaard supplemented by still-unread articles from *Science*, *Nature*, the Danish paper *Weekendavisen* and the periodical *Ingeniøren*. Every journey has commenced with a book on the plane, but hardly a single day on safari has passed without me opening a book to determine the identity of a green beetle, a pretty blue flower and a downy, young bird of prey. Many a midday have I relaxed with a thick book in the heat just across from the

Mara's lazy hippos. This has often been for pure diversion and frequently in a determined attempt to clarify what I've actually seen.

Life in literature and life in the wild are often two sides of the same coin, and I am far from certain from which comes the greater inspiration. I freely admit that I have drawn enormous inspiration from my great heroes, Richard Dawkins, Jared Diamond, Stephen Jay Gould, Edward O. Wilson and Steve Jones, without whom I could hardly have put a thing down on paper. The books and authors to which I constantly return are however of a completely different stamp: Gregory Bateson, Søren Kierkegaard, Michel Serres and François Jacob besides, naturally, Darwin himself. Where one group is concerned with nature, the other is preoccupied with the 'spirit' in nature. And there you have my odyssey: if I may permit myself a moment of solemnity – a fusion of theory and practice, ancient and modern, spirit and matter. Light and dark.

Most people are familiar with the famous saying of Socrates, 'The only thing I know is that I know nothing." At that point in time he had already demonstrated a monumental wisdom, yet managed to add that he was in fact so wise that he understood the extent of his lack of wisdom. Perhaps he was so wise that he also had an idea of everything he did not even know, about which he was ignorant.

Once, however, we knew that the world was flat. Of course it was, otherwise we would fall off. How stupid is one permitted to be! Today we know that the Earth is round, and that the force of gravity prevents us from falling off, but the Earth's roundness is limited to the three dimensions, and thanks to Einstein we have an idea that one should reckon on a fourth: Time. Is the Earth still round? We have some very vague ideas that there exist many dimensions, possibly infinitely many, but we have no idea at all what this entails. Everybody is familiar with the word 'infinite', but do we have the least idea what infinity is?

In this respect nothing has changed since Socrates. In those days most people were already aware that the Earth was round. Aristotle had 'figured' it out. All new knowledge simply reveals that the mountain is higher than it was supposed to be and that there is always another, higher ridge behind.

It is in our nature to be inquisitive. Perhaps we have an especially inquiring mind, perhaps it began to be inquisitive once Eve ate the fruit of the Tree of Knowledge. We seek wisdom about life and call it the truth. Fortunately the philosopher Nietzsche has taught us that simply to seek within oneself is life-promoting, and fortunate this is indeed, since there is something of a Labour of Sisyphus in which we have become embroiled. Sisyphus was the unfortunate bloke who had been doomed by the gods to roll the stone up a mountainside forever. Every time he neared his goal, the stone would be tipped down again. We seem doomed to die in uncertainty. 'Ordinary life' is simply not long enough, but perhaps life is the correct length *in any case*.

We can choose to lie down and die, we can let ourselves be entertained by *X-Factor* or golf or continue our inquisitive search because that in itself is life-promoting.

Let's begin at the beginning, and in the beginning was Africa, as is written by St John the Evangelist.

ABC of reading

The American poet Ezra Pound was an inordinately learned man, who is regarded as the style-setter among the great poets of the twentieth century. His texts, however, are so difficult that he supplemented them with a textbook on reading comprehension with a concluding answers key. An ABC of Reading.

I myself nurse a humble hope that this book can stand alone and will therefore be content with a fairly brief introduction to the following pages where I reflect on, and express my astonishment at the way life has unfolded and developed from a poisonous bacterium to a poet who writes difficult poems about life. I can, however, reveal that what lies ahead are more questions than answers.

"Science has demonstrated" ... no, "As far as we know," the first primitive live arose on Earth some three to four billion years ago as extraordinarily simple, single-celled bacteria, which, in a fairly short time, developed with notable success. Like a closely fitting membrane, a biosphere, life wrapped itself

around the inorganic, lifeless Earth. This worked really well, so well in fact that the first organisms ended up polluting the entire globe with their waste product: oxygen. Soon lethal oxygen covered the biosphere and threatened to kill everything, but life refused to be blotted out. Many organisms moved down into the oxygen-free layer in the crevices and cavities of the Earth, while other pioneer bacteria mutated and gradually accustomed themselves to living with the poison, which they also learned to consume. The oxygen could suddenly be used, and a new 'oxygen motor' kick-started evolution, which produced plants, reptiles, birds and mammals, including people. The human being became so successful that today it has to take up arms against its own waste products.

This book deals with the mechanisms that propelled life here and there among numerous fortuitous forms, at least one of which is capable of reflecting on its own role in the great network of organisms. We humans have acquired the facility of feeling we are something special, and thus also the feeling that we constitute an evolutionary zenith. That we are too, although only in our own consciousness. It takes a human to understand a human as something special. Nature is indifferent. Here there is only one criterion for success, namely survival. What the brain can produce in terms of poetry and nuclear physics is beneath notice compared to the ability to survive. If we accept the prehistoric people *Homo habilis* and *Homo erectus* as the first human beings on Earth, bacteria are still thousands of times older and are currently the most successful organism.

A warthog suckles her young.

We human beings differentiate sharply between these organisms, between people and bacteria, Nevertheless, our organism harbours many billions of bacteria, divided up into thousands of species. Our cells and bacteria communicate with each other, however, and, working together, accomplish a number of vital tasks. Without this commonwealth there would exist neither people, brains nor poems. How can we establish the boundaries of a person? And in relation to what?

The book is divided into smaller sections, each with a superscription and, as a general rule, a concluding story or point concerned with the development of the organisms. Most of the sections introduce tracks, themes and concepts which are expounded later, and the story continues from section to section down many crooked pathways. To facilitate an overview I have brought the sections together in larger groups, each with an overall superscription.

The account of the mechanisms of evolution begins with our still-living forefathers in Africa, where the conditions which shape the human being's development and thought process are easy to spot. All life is shaped by the fight to survive and shape oneself, as if taken directly from Maslow's Hierarchy of Needs. We must have food and water, we must mate and protect ourselves

from our enemies. These essential needs also direct the introductory section where I have chosen to take the point of departure in the relation of organisms to life-giving water.

We call the organisms' gradual or continuous adaptation evolution and we devote innumerable thoughts to the character of the controlling mechanism. This discussion is several hundred years old and is absolutely the pivotal point in the following sections, in which I represent a part of the history of ideas and some of the most important points of contention. I also expound my own viewpoint that pervades the entire book, namely that the development of all organs and organisms takes place entirely fortuitously in accordance with purely rational principles.

In the process the various survival strategies of the organisms are introduced; these can be roughly divided up into statically conservative and dynamically developing – or, put another way, with and without a brain. This is a thread of scarlet which runs all the way through the book.

Following the discussion of evolution the elephant is introduced in the section 'Cogito ergo sum' in which I expound on the senses of animals and humans. I regard the senses as one combined quality, the sum of which always has the same value. These are mutually developed and optimised. What is lost in on the roundabouts is gained on the swings.

After this comes the communication track that runs throughout the rest of the book. I examine and discuss communication in other mammals and plunge into the history of development and find communicative traits in the world of plants, in micro-organisms and back and forth across the species. On closer examination communication appears to be present everywhere among the Earth's organisms, and one easily falls prone to the semi-religious thought that everything that lives is in mutual contact. This suggests that holistic thinkers such as Gregory Bateson and James Lovelock have a point when they insist that the whole is greater than the sum of its individual elements; that one does not perceive the full picture solely by studying the details. Holistic thinking is a break with the established scientific thought process, which tends to reduce everything to the smallest units. I have no ultimate answer as to which is right, but why limit one's options?

The last and most controversial main track deals with feelings, which here become the evolutionary glue that binds certain animals together in social systems. The main argument is that evolution always recycles the good ideas. When and if feelings are developed, they must, of necessity, appear in many contexts. Is it important that feelings have their roots in chemistry, and is there any difference between a human being and an elephant in this context?

I permit myself a certain historical manoeuvrability, leaping forward and backward on the timeline, until I return to the development of the human being at the end of the book. Before that I have some thoughts on social tendencies, nature's network and co-existence together with the genetic mechanisms that may carry some organisms forward towards self-awareness and emotions. I propose the possibility that certain evolutionary processes move much faster than we realise.

Finally, I am ready for the story of apes, ape-men and human humans, together with the crucial conditions which have influenced our development. There is, unfortunately, much that we do not know – in particular, we do not know what we have become!

Søren Rasmussen
Holte, Denmark
January 2011

Water thick-knee.

Old harbour at Zanzibar.

Baobab. BAOBAB. The sound has a tang of Africa, in the same manner as a blend of red earth, the Masai and elephant dung. In Africa the senses merge: they are no longer to be trusted. Baobab. The mighty tree reaches up high into the deep, blue sky, in which the uppermost twig fades and becomes one with the cosmos. Life streams in through the brown tissue paper of the bark sheathing the powerful roots until they grow to monumental dimensions, proliferating as a riotous network in the underworld. Baobab, Africa, creation, Bushmen, roots and senses.

The Bushman myth says that the tree stands on its head, but the pumpkin-sized fruit on the bare branches give the lie to the myth, which cannot, of course, withstand rational analysis: it's not meant to, since the Bushmen make no distinction between belief, knowledge and a good story. The tale of the naked baobab is ancient, dating back to the time when people got through a greater number of gods. It tells how one god became angry with the people, because then, too, people built towers of Babel. As a punishment he turned all the trees in the world on their heads. Fortunately, the god was of a merciful and forgiving variety who had compassion on the people wailing in distress, and repented of the misfortune he had brought on them. "Let that be a warning!" he said, and turned the trees back again so that they stood with their leaves towards the sky, just the way trees should. There remains Africa's fabled symbol, the mighty baobab tree, which, in his haste, the god forgot.

I am peeling an orange in the shade of the tree and retelling the myth to my travelling companions, who, not without a hint of enlightened European arrogance, are amused at the childlike naiveté of the Africans, just as Karen Blixen and other Europeans were in former times. The Bushmen are, of course, quite aware that the tree does not grow downward, but the myth serves to remind them that trees and plants are living things, and that everything living is interconnected and forms part of the same cycle with the same status. People cannot live without trees, but the trees seem to do amazingly well without people. Living things are made for each other, and if people forget this it might just be that the trees will be inverted again.

We walk round the tree, reckoning its circumference at twenty-five metres. The age is difficult to determine, but the size indicates that the tree is probably more than 1,200 years old. This being the case, it is among neither the oldest nor the largest of the species, but its spores lead right back to the Early Middle Ages; to the time before Saint Ansgar brought Christianity to the Danish Vikings; a long time before the people whose forefathers originally left Africa, returned in a somewhat paler version as colonial overlords.

Year after year, generations of elephants have sought refuge in the shade of the baobab tree, bored holes in the thick bark, sucked thousands of litres of water from the tree and requited with a portion of manure. Year after year the tree has provided water and food both to people and wildlife and has dutifully healed its extensive injuries from elephants' tusks. And it has grown and grown.

The old tree bears the brutal scars with Nature's rugged grace. Its history is chiselled into all its furrows and wrinkles like the face of a Caucasian centenarian, who has equally dutifully eaten his home-made yoghurt his entire life. And life goes on.

In Selous National Park in Tanzania, the giants seem randomly scattered, as though by a gentle hand. Somewhere on the horizon, the naked silhouette constantly towers over elephants and acacias. Perhaps the oldest contain a history of 4,000 years, which we can only partly wrest from them. For what remains we will simply have to have recourse to imagination.

Exhilarated by the African atmosphere, I cheerfully interweave the interesting facts about this extraordinary tree with local mythology and superstition, for who defines which reaches highest, a baobab or a bird call? And so we say absolutely nothing about relativity, but a grinding sound, already of high volume, attracts my attention.

I don't manage to look upwards before the small remains of a jaw which evolution has left behind in my middle ear, send the sound farther along the acoustic nerve to the brain, the database of which informs me that one of my distant, feathered relatives is seeking contact from the phylogenetic baobab.

Honeyguide Camp (also facing): beautiful, traditional, first-class bush lodge.

This thought is thought through to its conclusion, long before the eyes focus on a reticent, greyish-brown bird about the size of a starling. A honeyguide. If the thought is swift, then what about the creation of the universe?

The tail flares out like a Chinese fan, and the sound strikes one as far too great for such a small creature. In the same moment that I get up and move closer, it shoots up like a jack-in-a-box, chatters on, flaps in a quite exaggerated fashion, proudly fluffing himself up, as that old teller of fairytales, Hans Christian Andersen, would doubtless have expressed it. It flies twenty metres to the nearest tree and then the whole performance starts all over again. The little honeyguide wants to lead us to the nearest beehive, and dives at the nest in expectation that we will crush the hard exterior and get the sweet-fixated bird access to the glorious contents: the wax – the honey it leaves to us.

The bird is a creature that loves the easy life, and who, in cuckoo-fashion, leaves the hatching and care of its young to its small, humbler colleagues while it entices the human being to provide it with food. No mean achievement in a world where man has to earn his living by the sweat of his brow and has seriously taken the lead over all life-forms.

The honeyguide is the only wild creature on the planet which quite sha-melessly makes use of man's presence in the neighbourhood. How long this collaboration has been going on, nobody really knows. But birds are feathered lizards, an ancient creation which attained its form many millions of years before apes began to walk on two legs, and long before they developed the idea of being chosen from among the creatures of the Earth. There are, therefo-

The story of evolution seen from the African savannah

re, all possible grounds for believing that it happened long before modern man made his entry into evolution. We probably have to go back at least 4–5 million years to our ancestors 'Handy- man' *(Homo habilis)* and the 'Upright Man' *(Homo erectus)* and even farther back to their ape-like ancestors. They certainly crushed bees' nests and gulped the sweet honey down following the guidance of the little bird.

As a crucial precondition for this collaboration, the honeyguide developed a unique enzyme, which can break down and digest the otherwise indigestible beeswax. In fact Nature does not permit the existence of non-degradable substances. A single mutation in the bird's genome has converted a particular protein enzyme and thus created a new possibility. This is what we call evolution. Other mutations have curled the human brain so we can give ourselves the strangest fancies and, for example, conceive of the bird as a social construction. How far this is wide of the mark or a brilliant opportunity, I shall leave unsaid.

The honeyguide has incorporated its curious behaviour into its genes, which is hardly anything exceptional in Nature. It does not have to learn its behaviour every time a new generation takes flight. Its genome remembers from generation to generation that it has to cooperate with man and the code of the protein enzyme which carries out the digestive process. It has doubtless taken a considerable time to develop, just as we will speculate for a considerable time on how it actually came about. And what else is on our gene map? What kind of behaviour is in our genes? Something particularly human?

Man's alliance with the honeyguide is something the Bushmen learn at their mother's knee, as it is with the other people of the savannah. They have prized it for thousands of years just as they have learned that the cavity in the baobab tree is a natural place for an enterprising swarm of bees to congregate. When the baobab tree was a runt for a few hundred years, the Bushmen began to cut steps in the trunk to facilitate access to the sweet honey. Even so, the skilled hunters of the savannah could never find a way to cheat the bird of its fee. The Bushman legend relates that one must always show the honeyguide one's gratitude by leaving out plenty of food for it, or punishment will follow on the next honey hunt when the arrogant will be led to a dangerous, venomous snake or some kind of predator.

The Bushmen have lived side by side with the honeyguide in southern Africa ever since modern man first saw the light of day. They are the eldest unchanged representatives of modern, thinking man. Before us and the Bushmen there were many other kinds of humans and some of these lived parallel with us, but the exact relationship with these is still unknown.

The Bushmen broke out of East Africa and moved south, a long time before their relatives moved north and jumped over the Red Sea, out of Africa. For more than 100,000 years they have lived in southern Africa's dry, sunny surroundings and developed a perfected culture which is in no way less advanced than other cultures of the world. They have not built pyramids, railway bridges or flat-screens, but they have lived well and produced sufficient reserves to facilitate companionship, celebration, song and risqué jokes. They have developed a culture in which children have no duties but play while the children of civilisation continue to tread the pedals of sewing machines in foreign textile factories.

The Bushmen are small, agile people who can live extremely frugally. They have a capability of building up fat deposits around the thighs and hips, which is known in technical parlance as steatopygia. This functions almost the same way as the hump on a camel, as reserve layers which can be transformed into water and energy in times of scarcity. In all probability steatopygia is a fundamental human capability, instances of which can still be found scattered all over the globe, while with most of us 'emigrants' the capability has temporarily disappeared.

The small Bushmen seem to represent a formidable adaptation to some of the planet's most inhospitable regions. They have learned how to do without food and water for days on end, while the body shrinks and begins to resemble dried apples, until the skin smoothes out again under more favourable circumstances. It is almost like pouring water on a shrunken succulent. Under these conditions, their endurance and the ability to produce layers of fat give this section of the human race the possibility of optimising their living conditions.

Historically, the Bushmen have been described as primitive, and when we do acknowledge the kinship we are inclined to regard ourselves as a kind of modern subsequent development. Nevertheless these people possess what, in relation to the languages of the modern world, is the most complex language

Entabeni Mountains offer fantastic walks.

on the planet, with five extra sounds (phonemes). These extra sounds are pronounced as clicks with the tongue against the palate and are a part of their highly perfected and detailed description of all natural phenomena: every leaf, grain of sand, function and nuance has its own sound. They know all the sources of food and can eat, generally speaking, whatever is to be found in a desert, from insects, seeds and plants to mammals, bark and fruit. One knows what one is talking about.

The first child usually comes somewhat late compared with other 'people of Nature' and then the children are born at four-year intervals. The body regulates itself: it cannot manage the production of mother's milk more frequently in these frugal surroundings. There is a close connection between the size of the population and the bearing capacity of Nature. The Bushmen do not produce more children than Nature can provide with food. For this reason it is a stupid myth when white South Africans claim that the Europeans took over an empty country in South Africa. The country was full to bursting with exactly the number of people and animals that could be accommodated.

For many thousand years the Bushmen had no metal – but then they did not know they had a use for it either – before the Europeans sold them their old barrel hoops. In the meantime they developed everything they needed from the materials to hand.

More than 5,000 years ago they developed a refined, high-technological hunting implement: a small bow, easy to handle for small people moving fast. They learned to sneak up on even large animals, run close and fire a small, deadly arrow with a refined poison. The arrowhead is a very sharply pointed ostrich bone that will penetrate even the thickest hide and deep into the bloodstream where the deadly poison starts to spread. The arrow is made of several sections so that the actual poisoned tip remains in the wound and continues to work, even if the prey has broken and ground off the shaft. Inuits use the same technique with harpoons, where the spearhead is attached to the harpoon line, with the difference that there is no use of poison.

The Bushmen have devised a wide variety of poisons from plants and grubs, which, in their application seem so thoroughly developed that they stand comparison with equivalent high-tech products. For example, one plant poison consists of strophantine, a cardiological glucoside which slows down the heartbeat of the victim which then is run to exhaustion. By using the correct dosage of 'heart medicine' it is possible to avoid ruining the meat, and should one be unlucky enough to poison oneself there is help at hand from the seed of the baobab tree, which contains the alkaloid adenosine, an excellent antidote.

The Bushmen have reconnoitred and investigated their natural environment to the minutest detail. They have investigated the application possibilities of all animals and plants and developed innumerable complex solutions which are hardly even surpassed in modern societies when seen in relation to the environment and its possibilities. At the Institute of Bushman Studies in Windhoek, Namibia, efforts are being made to preserve as much of this knowledge as possible.

Bushmen, Danes, Chinese and Jordanians belong in every sense to the same species, the human being, or, more specifically, modern man or *Homo sapiens*, as we say to mark the difference between ourselves and others, i.e. earlier species of man. In principle there is no difference between a Bushman eating a delicious grub or a raw ostrich egg for breakfast, and an Englishman enjoying egg and bacon fresh from the frying pan. Even so, we believe there is a difference, which we have the temerity to call development and civilisation, but is a fried egg really better than a raw egg? Our notion of development is as something positive, progressive, which creates new possibilities of a primarily material character. We only associate the concept of development with the people who left Africa, but it may well be permissible to inquire how the development has enriched civilised life with us emigrants in relation to the materially frugal life of the Bushmen of the Kalahari and its surrounding regions.

Arrow-marked barbler.

WHITE MAN IN AFRICA

Experience:
when you know
that what you have learned
and know
is actually correct

Bloodbrothers

Nairobi, Kenya.

Karen Blixen's house with the famous view of the Ngong Hills has achieved the status of a must-see for visitors and, indeed, has been incorporated in Kenya's National Museum. The farmhouse has been restored and attractively furnished with pieces of the sets from the film *Out of Africa*, by which one shouldn't let oneself be put off, as the result is amazingly close to the original which can be studied in numerous pictures. The atmosphere exudes upper-class life in Africa: safari, erotic adventures, culture and great literature.

It didn't end too well with the hero's death, bankruptcy and flight, which is the stuff of good stories. I find it difficult, however, to accustom myself to Karen Blixen's many names, e.g. Isak Dinesen and Pierre Andrezél besides the American synonym Meryl Streep.

The way to Blixen's famous 'African farm' is called Langata Road, one among many African roads marked out by use. According to Blixen, heavily laden ox-carts and assorted equipages with lords and baronesses left deep wheel-ruts in the red African earth, while black servants and colourful tribal folk wandered between the animals. And thus a road was made.

Cast a glance behind the inhospitable, barb-wired walls and iron gates along

24

the road and one discovers however, beautiful gardens and parks, where thousands of African gardeners have their daily occupation. Long boulevards cut through the hidden landscapes and lead to an expanse of gravel with large English bungalows, servants' quarters, stables plus a tennis court and a pool. A generation after the achievement of Kenyan independence, the country houses are still door to door or gate to gate in the suburb Karen. Lovely it is, the white town, where the small dwellings are for the African chauffeurs, gardeners and porters: 'upstairs and downstairs', but not everything is black and white.

Today many Africans have taken their rightful place among the white families and can enjoy high tea and a nip of gin and tonic on the terrace. In Karen Country and Golf Club you see more and more African hands swinging clubs, even if one is still spared the experience of white hands washing balls. In Nairobi's better restaurants the Africans are gradually becoming the majority. Here they make a virtue of the two-hour lunch, preferably followed by a well-earned siesta.

Equestrian sport is the unmistakable imprint of the British Empire in every respectable former colony. Polo and racing enjoy just as high a standing with the new whites as cricket among the old whites and rugby with everybody else. And Nairobi's race course is naturally situated in Karen where women continue to be invited to dance among the town's prosperous blacks, whites, yellows and reds. A hot Derby Sunday will see a swarm of black hands bravely waving the day's white programme, and one is left in utter doubt as to who is now a real 'African'; i.e. Afro-African or Euro-African. It should be mentioned at this point that it is possible to dine really well at the course's Racing Club, which I, with great enjoyment, have frequented on numerous occasions. One evening I visited the establishment together with my old quick-witted friend and safari guide Jamie, one of those second-generation immigrants left behind by the empire.

Jamie belongs to the lost generation of children and young whites without a family fortune who fell straight down into the no-man's land between the British withdrawal after the Mau Mau uprising and literal African independence, when many whites were suddenly a tolerated presence unless they succeeded in being promoted to a 'necessary evil'. The girls moved out or married out of the problems, while the young men formed a small, closed group of dare-devil Kenya cowboys, who supported themselves as safari guides, diamond smugglers, bartenders, all-round handymen and con artists.

Jamie McLeod was always full of stories about dangerous, adventurous safari life, which I swallowed raw; until I was confronted by a couple of my own with new actors. Here, however, is one of the better ones.

Jamie was a small guy, and like many of the whites he grew up with horses, exactly the combination that makes a Derby winner. And he won the Derby.

The performance of his life, and an achievement which could not be greater in the Kenya of the past. As the white champion of the young nation he was presented with the trophy by the Father of the Country, *Mzee* Jomo Kenyatta himself, who was, however, a touch resentful at the many whites who were still in the saddle.

Kenyatta vented his irritation on the unoffending winner with a remark about Africa for the Africans. Jamie, always ready with the repartee, replied that a little scratch in their thin skins would make them both red. It got awfully quiet around the Father of the Country, who was known as a man one didn't contradict. The silence was all pervasive, but Jamie was untouchable in that great moment and just went right on, saying that he was a Kenyan, born in Kenya, and that it was not his fault that he was not a citizen in his own country. Kenyatta concluded the encounter with a movement of the hand, and Jamie was instructed to meet the great man the coming Monday at midday. Cocky as ever, he showed up at the president's office, and what transpired there we only have from the horse's mouth, so let us be content with stating that Jamie received his passport from Jomo Kenyatta and was thereby accepted as a citizen.

For a number of years Jamie was a close colleague, safari guide and camp manager. Like most other Kenya cowboys he had a long and exciting life, even

if somewhat more concentrated than average. A Minnesota Cure in England, paid for by friends, gave him an extra five good years, but a daily diet of three bottles of cheap vodka and 100 appalling Sportsman cigarettes hastened his demise. Honour to his memory.

Bonfire dialectic

For a brief moment the embers flare intensely in the imperceptible breeze. The light accelerates in the black hole, the mass expands. A moment later the charred black parties are back again. A split-second, like the history of mankind on Earth. The brief moment from when we landed on two legs until we began to think about black holes.

It's a real fire of the good African kind with a half-metre-high, skittle-shaped mass of embers and radiating heat that can be felt many metres away. The heat is too extreme for making tea or frying, so I shovel a good portion of embers out for the purpose. As I like making fires and Mohammed is going to make tea, while Jamie prefers to watch, the division of labour is in place. I have to straddle the thick stump and use both hands to lift the next bit into the embers. That's the way we make fires in Africa. It won't win any boy-scout points but it works.

It is only fourteen hours since I landed in Nairobi. Jomo Kenyatta Airport

Kilwa Island, Tanzania: Arabs settled on the Swahiili coast more than 1,000 years ago, controlling the lucrative trade in African people. They were followed by the Portuguese who built a number of fortresses along the coastline.

hasn't changed much since my first arrival with the old Kenya Airways 707 at the beginning of the 1980s. The whole thing still feels slightly scruffily African but my original insecurity in the face of the unknown, black continent has today turned into a warm pleasure of return and a never-failing sense of tension and adventure.

East Africa is the cradle of humanity. Here human-like creatures developed one after another, and it was from here that the Earth was populated. Here we learned to walk on two legs; here is where our brain developed, and, I daresay, here is where we learned to use it. From here a small, tough group of thinking people set out and populated the rest of the world. This is where all of the animals lived which our forefathers hunted, and by whom they were hunted. The stage is intact today. I cannot conceive of any place having a greater power of attraction over people.

Mohammed's Arab *gallabya* is easy to spot in the bustle of the airport. He is the only one in authorised Muslim dress. I take pains to remember the sequence of traditional verbal courtesies in Swahili with a dash of extra Arabic colour. It would be disappointing if I had already forgotten them. We follow the beaten path along Nairobi National Park towards Uhuru – the Freedom

Highway – towards Nairobi and beyond. On the way to our primitive camp in the Masai Mara we talk about everything under the sun. Mohammed is a chauffeur-guide and is eager to hear all my stories about animals with which he will later divert all his guests. I tell the story about the honeyguide and man; he knows the story but also wants to incorporate the honey badger, a nasty little creature from which one should keep one's distance. According to Mohammed, the honeyguide also leads the badger to the bee-nest. So far this has never been proved and this is probably one of the many folk tales in which Africa abounds. Instead I share my reflections on the honeyguide's collaboration with man. I sense that we have contrasting views on the subject, but, on the other hand, Mohammed has no desire to contradict Bwana Khupa – the big boss – which is my 'secret' nickname.

The embers of the fire, the starry sky and the sounds of the night encourage conversation. When we have guests at the camp Jamie entertains with a whole series of tall stories, some of which are more plausible than others. When we're alone he grows more thoughtful, stares at the fire while clinging to his tea mug. I know that he's listening intensely while I tell Mohammed and the other African employees my own tall stories which are borrowed, but not actually pirated.

In the 1980s research approached a consensus on man's most recent developmental history and it was clear that the entire present population of the world came originally from Africa. Chinese, Aboriginals and Europeans thus did not each emerge as a result of a long, independent development, but are really closely related cousins who emigrated from the dark continent relatively recently.

It has been genuine solace to me to take every opportunity to tell Africans I encounter that every white person is in reality a pale version of an African. Neither can it be denied that I'm piqued by a malicious, inner joy while simultaneously thinking of how this development is received by the stupid racists all over the world. Mohammed and his colleagues are being sedulously amused with the scientifically acknowledged 'tales' about the little African tribe that left the African continent about 60,000 years ago and populated the rest of the globe, while the outer features slowly adapted to suit their new surroundings.

As a rule the story calls forth a series of typically high peals of good-natured laughter. The first time because it is conceived of as a joke. The second time because it occurs to them that I mean it seriously (I can't be that bright) and the third time when they understand that the story is quite possibly true. I can still sense that this thought has just as difficult a time taking root in Africa as in the American Bible Belt. Nevertheless it is a story that science does not hesitate to call 'evident', which sought to be one of the finest.

Mohammed has heard of Darwin, which is sort of synonymous with the discussion over how far we are descended from the apes. Jamie listens from

Fishermen on the East African coast.

the sideline, but only contributes with a recurrent shake of the head. In reality they want to hear what I have to say on the question and hear my crazy arguments, for they know all too well that we are not apes. So I decide on a different approach with the question: "Are we all Chinese?"

No! We are certainly not Chinese; thus far we have a consensus! For the first because we do not look like Chinese' our eyes are not slanted, and our skin is not yellow. On further consideration we would add criteria such as language and culture in this evaluation. We are quickly agreed that it is possible to be a Dane and resemble a Chinese, or Chinese and resemble a Dane or an African.

These criteria are fluid. In most cultures there is a general acceptance of a person who speaks the language and has absorbed large parts of its history, religion and mode of existence, even though the appearance can diverge substantially. Chinese, Africans and Danes belong to the category we call human beings, a category we have defined from biological criteria. One is the same species when one can have procreative offspring.

So then I tighten the screw: the distinguishing of people from animals is not a natural phenomenon, but a barrier which man has erected and defined, and which we constantly try to redefine and underpin. This is where the audience get up and leave. A human being is a human being, an ape is an ape and an animal is an animal. And that's it. Jamie's head-shaking assumes a partly insulting character.

I inquire rhetorically if it is not interesting to discuss differences and similarities; to try to establish what are human characteristics? Do we understand why we are so special? I reiterate that it is only in our eyes that we are something special. Now it's not going to be possible to discuss this question with the elephant, but I am, however, fairly confident that it doesn't see anything particularly outstanding about the human being. Certainly not in relation to a fellow elephant.

Mohammed believes that God created us and made us special. That is beyond discussion. In fact any discussion of the point should be forbidden! We have surely crossed the border with blasphemy. Mohammed and the other African guides will gladly hear about Darwin, the giraffe's long neck and the elephant's trunk, as long as it doesn't end in Kipling's literary fantasies. They understand the logic in the principle of development, but belief keeps reappearing all the time and catches them in a dilemma between the logic they immediately perceive and correct belief. As long as I stick to nature I can offer them everything, but the moment I bring man into the discussion it all goes completely pear-shaped.

Therefore I have to declare myself in agreement that the human being is in a category of its own, if only for the reason that there can be a need to establish which species we are talking about, just as it may be significant how far we are discussing an African or a Chinese in a given situation. However, if we define our own species or category against the background of established criteria – apart, that is, from creation, so as not to be guilty of blasphemy – we must ourselves determine whether apes or other animals have so many characteristics in common with us that we are prepared to let them into our own category.

Thus the question arises: what else shall we do with this integration? If we determine that human beings are apes, we will probably maintain that apes have various categories – tribes, species etc. What is the purpose of this exercise? If we believe that apes are human, is it because we judge precisely that those characteristics and values which make us human are to be found in such abundance in apes, that we have to regard them in the same way as we do ourselves? In practice it means that they are entitled to some kind of 'human rights', which begs a new question: how does one safeguard the rights of individuals with whom one is unable to negotiate? And then which other animals are entitled to corresponding rights?

Now neither Mohammed nor Jamie can envisage any especially human qualities in apes, so the answer is quite simple. At this point in time we were unaware that the future would bring about a political discussion on the human rights of apes.

Black or white?

The temptation to conduct oneself momentarily in clumsy Swahili can be hard to resist: the simple ordering of a couple of beers accompanied by a convincing proviso that, if possible, they should be extra cold and concluding with a courtesy worthy of a king. Mostly because I can! The answer will, with unerring certainty, be given in eloquent, impeccable, polite English, clearly demonstrating that the

African in question communicates in quite excellent English and that, moreover, it would be an extremely good idea if I did the same.

Mohammed, Peter and Rafael teach the local languages with great enthusiasm, and, gradually, Swahili, Maa and Kikuyo blend into a hotchpotch, which, however, is not carte blanche to show off. If I give a message in anything other than English in front of our guests, I am most frequently answered with long, rapid, completely incomprehensible sentences. So as not to appear a gaping fool I have developed a suitable reaction: a meaningless grunt, very reminiscent of that of a gnu, and leaving the impression that I am completely up to speed.

Distance, barrier and prejudice are embedded in a hundred years of colonial history, which comes across more clearly in numerous jokes. One characteristic African story is about a white man who thinks his Swahili is pretty good. At the dinner table he raises his bent fork and asks in Swahili if he can have a decent one. Unfortunately for him there is only a modest 'k' in the difference between the word for fork and the name of the female sexual organs. So anyone can work out the rest. This story is a sure-fire winner in Africa where it always makes people laugh until they cry. Most of the stories about the white man feature people with a tendency to think too much of themselves.

The European 'Wise Men of Gotham' stories about Africans are very often about events of which the teller has personal experience, so that the authenticity seems beyond any doubt. I retell the story of the African cook who was supposed to serve a sucking pig, roasted whole in the European style with an apple in its mouth. For want of an apple he was told to serve the pig with an orange in his mouth, which resulted in him bringing the roasting dish to the table with an orange in his own mouth. Every African finds this story insanely funny and one is probably convinced that it is correctly told, for yes, there are many stupid people, and so one does not reflect that the transmitter of the story may associate it with a particular skin colour.

One can hear the sighs in the whites' stories of stupid Africans, but this is a burden one takes upon oneself. The stories contribute to the widespread myth which says it was the smart people who left Africa. The myth about the white people who are quicker, cleverer and more innovative than others gradually crumbles away as it occurs to us that the races of the planet have the same close point of departure and share an ancestral mother and father, who are surprisingly young. The total population of the world are cousins irrespective of their skin colour and nationality, and the emigrants were unquestionably black because the genetic changes which have brought about the pale skin colour are far younger than the emigration.

The weird thing is that white people have apparently always had an inclination to put a distance between themselves and other alien or exotic people at the same time as they become conscious of them. Thus Columbus

Above: Bushmen of the Kalahari.
Below: Honeyguide.

Above: In the beginning was Africa ...
Below: Election day.

Above: Hadzabe hunters in a baobab 'forest'.
Below: The author exchanges jokes with Masai friends.

Homo habilis, taken from Sterkfontein, South Africa.

Above: Zebra take a nervous moment to quench their thirst at a waterhole.
Below: Crocodiles swarm and gorge during migration season.
Bottom: A dried-up waterhole offers little solace to this hippo.

Above and right: A dangerous run: zebra and wildebeest brave river crossings on their annual migrations between the Serengeti and the Masai Mara.

Wildebeest and impala are staple prey for lions, but there are no free lunches in the African bush.

The legendary white lions of South Africa are finding their way onto commercial breeding programmes.

had hardly set foot in the western hemisphere before learned Europeans were having recourse to Aristotle's arguments about the lower status of Africans in order to enslave the godless, dark-hued beings on the other side of the Atlantic. The learned conceived of the whole world and every living thing as ordered in ranks with white people, or rather the white man, at the top. God had created the white human being above all other creatures and, as a result, the latter had to bow their heads in submission.

At this time the Portuguese were doing well in the African slave trade, without any debate worth the name about the status of the African slaves, which apparently, without further ado, could be left to the inhabitants of the New World. Since, however, the appearance of the Indians was closer to that of whites than the Africans, the world's first debate was about whether foreign people actually had the status of human beings. How did one establish a boundary between humans and animals? It was obviously not as crystal clear as had otherwise been imagined. A human being was, after all, a human being, wasn't he? What criteria should a human being meet? Colour of skin and appearance? Belief in God? And no less important: could a human being be a slave? If it was accepted that a very black African could be a slave, what of the light-skinned Indian? There seems to be a certain consensus that the ungodly had no rights: they could be deprived of their property, enslaved and even murdered. What if they agreed to convert?

Bartolomé de Las Casas came, like so many other young people, to Mexico in 1502 to seek his fortune in the new Spanish possessions, but, after a few years he ended up distancing himself from the extreme, lethal exploitation of the local population. Initially he believed that slave labour should be carried out by the sturdier Africans, but later he entered the Dominican Order and dissociated himself from all kinds of slavery.

Las Casas was far from alone. There was a thin but steady trickle of Catholic monks, for here there was work to do among all the heathen, whose language and culture they gradually appropriated. The learned monks were, to put it mildly, surprised at the cultural level of the Indians, whose family organisation and children's upbringing were far superior to those of the Spanish, as was noted by Bernardino Sahagún in *The Florentine Codex* the most famous book of the age.

In 1537 Las Casas had a papal decree, *Sublimis Deus*, drawn up which establis-hed that creatures in the New World were rational, thinking beings with souls, whose life and property should be protected. The colonists were, however, not so easily won over, and Las Casas had to go through a lengthy debate in Val-ladolid presided over by the Spanish king Charles I, also known as Charles V, Holy Roman Emperor. The clergy at the debate voted for Las Casas; the Indians were people, and people could not be slaves. The judgement, however,

A cheetah, the fastest land animal. Private game farms and reserves do great work to help protected animals. Here the author's daughter is at work at Entabeni.

only covered inhabitants of the New World, not Africans.

Governments of Europe and intellectual elites had problems with the moral issue. Slavery was not consistent with the word of the Bible. Both the black people of Africa and the golden people of Latin America had sophisticated languages, family structures and religions. They were people, not slaves. In Latin America the problem was 'solved' by replacing slavery with a horrific feudal oppression, substantially no different from slavery, and the slave trade in other areas was justified on the basis that the Africans would thus 'ascend in faith'. In the book *Slavernes Kyst (The Slave Coast)* for instance, the Danish author Thorkild Hansen quotes the Danish bishop Erik Pontoppidan as saying. "… that the Negro, who is transferred from thence [from Africa] to the West Indies becomes far less wretched." As long as people were abducted and removed from the godless and wretched existence in Africa to a better existence as slaves with the opportunity

Above: A Mozambican woman.
Here two young Madagascan women wear traditional skin-protection masks

to sing hymns, things really couldn't be that bad.

The common people's ingrained concept of African people as of inferior status survived in the best of health right up to our day, and has, from time to time, spread into various layers of society and sects. When, in 1834, slavery was definitively abolished in all British colonies, it triggered the emigration of Boers in South Africa, often called 'The Great Trek'. The Boers left the English Cape Colony in protest against the Africans' new civil rights. The abolition of slavery was accepted under protest, but the spread of civil rights became the proverbial straw that broke the camel's back. The Boers were unable to live with Africans on the same level as Christians, in conflict with God's law prescribing a natural separation between race and religion, as they themselves expressed it. In newspapers of the time one can read that the Boers doubted that the African, 'even had a soul'.

Right up into the twentieth century much of the white South African population believed that black Africans were of an inferior race at a lower level of development. Their notion bound together appearance and shade of skin colour, while the cultural insight was, as a rule, absent.

Outward characteristics such as skin colour require only simple genetic changes and are some of the fastest and easiest to influence, as every breeder of pigeons or dogs has known since the days of Darwin.

Modern aquarium enthusiasts are able, without any training, to pick out the most colourful among the aquarium's otherwise grey, drab guppies, repeat the exercise a few times and so breed guppies with yard-long colourful tails which would be an invitation to certain death out in the natural world.

The genes which determine our appearance are rapidly affected and through natural selection we can change the appearance of humans in the course of a few thousand years. The pelt of an animal in captivity can even be changed from black to white in a mere ten generations.

Nothing, however, indicates closer links between the inherited characteristics that influence our appearance and other more deeply rooted attributes. Simply because one changes one's skin colour does not necessarily mean one becomes stupider or cleverer, kinder, faster or anything else. Other factors are necessary. Let us examine the truth incarnate about colours of skin.

The truth incarnate

Our godless cousins in America were not quite as godless as we thought; they had their own god, the Sun. Sun-worship was extraordinarily widespread in the so-called New World, where the inhabitants understood the connection between the sun and life. Without sun, no life and they feared that one day the sun would be extinguished.

The rays of the sun strike the Earth with a massive force that can kill every bacterium. Fortunately the distance between the Earth and the sun creates precisely the range of temperature and the strength of radiation which makes possible the existent life-forms, including the human being, just as long as we employ a certain caution in our dealings with it. We are a result of the influx of light and adapted to it in more than one sense.

The chlorophyll granules or chlorophyll in plants convert the sun's rays to growth, i.e. into the organic chemicals which also benefit animals and humans, for they cannot produce these substances themselves. We call this process photosynthesis, and it is the mother of all organic chemicals.

The evolution of the early mammals may well have taken place close to the equator, where the strong sun affected the cells of the body, which, on its side, has protected itself with hair, fur and pigmentation. Sun, skin cells and essential organs have gradually developed complex processes which can still be traced in the organism.

When the sun's rays fall on our skin cells, an important chemical is formed which we call Vitamin D. This 'sun-chemical' resembles the materials which the body transforms from food under the name of proteins, hormones and enzymes. We label these substances according to their functions, but they are all proteins which are produced in our cells, where their appearance and quantity are controlled by our genome – the genes. Vitamin D is the exception

Many Ethiopians are closely related to Europeans.

A Comoro woman wears a skin-protection mask.

which proves the rule: it is the only protein we know which the body cannot produce by breaking down and rebuilding food in obedience to orders issued by the genes. For that it needs the sun. Our genes deliver a parent material, a cholesterol, which is subsequently converted by sunlight into the active Vitamin D which then becomes part of our complicated metabolism on the same lines as all other proteins.

As long as the human being spends regular time in sunlight of a suitable wave-length, and, moreover, is equipped with a skin colour which both protects against overdosing and ensures the correct dose of sun, the outcome will be favourable. It is precisely these preconditions which were present during the early development of the modern human being, in which the human being was equipped with a dark skin under the burning sun of the equator. Not too much, and not too little: what we call natural balance. Our forefathers' dark skin thus offered protection from scorching and more serious damage The colour also restricted the receptivity to the rays which formed Vitamin D – but this was no problem as there was plenty of sun.

For every step dark-skinned people have moved away from the Earth's equatorial zone, their survival rate has been weakened, because the influx of sunlight became steadily weaker, so the skin colour ends up blocking the production of Vitamin D. Evolution, however, has partly remedied the problem through relatively simple mutations to produce lighter skin colours. The most extreme form, albino, is totally lacking in skin pigmentation. It doesn't have to be this extreme and small evolutionary changes towards lighter skin have been of benefit to the people who have moved farther north and south, away from the sun. Inhabitants of the north, in Scandinavia, have become completely light-skinned since all the dark-skinned people have fallen by the wayside en route. The survival rate of the dark-skinned has simply been too low, and the effect is ever more clearly visible, the farther north we go.

A completely light, typical Scandinavian skin multiplies by ten the body's capacity to produce Vitamin D compared to a completely dark skin, but, on the other hand, it offers no protection worth the name from the burning rays of the sun.

The various skin colours of the human being can thus be seen as a local geographical adaptation to the influx of sunlight. On the one hand we need to receive sufficiently large amounts of sun to produce the necessary amount of Vitamin D, and on the other hand we have to be protected against the most burning rays.

Light skin is, however, hardly a sufficient evolutionary solution when it's a matter of populating the regions furthest north. The sun is simply too low in the sky for most of the year, and the formation of vitamins is adequate for two months in the summer at most. Archaeological evidence indicates, however, that the problem is solved through diet. Fatty fish and sea mammals with masses of blubber contain, in fact, sufficient Vitamin D, in contrast to all other food. The Inuit of the Arctic probably first enjoyed real success when they learned to hunt sea mammals, and this may be the reason that the first immigration of musk-ox hunters in northern Greenland failed. They simply lacked vitamins and first had to find a food which could compensate for the missing sunlight.

The significance of Vitamin D is far from fully understood, but we know for certain that it is of crucial importance for the calcium metabolism in the bones. Lack of Vitamin D leads to soft bones and osteoporosis (brittleness of the bones) in both children and adults, a condition which was formerly fatal. We know also that Vitamin D features in a great number of metabolic processes, and that there is a marked connection between a lack of the vitamin and metabolic diseases, as well as certain widespread forms of cancer. These connections can probably be traced back to the importance of Vitamin D for immune defence, which is activated precisely because Vitamin D ensures that other transmitters can bind to our DNA.

There is historical evidence that the lack of Vitamin D has plagued mankind for a very long time. Thus the famous Roman doctor, Claudius Galenius, possibly better known simply as Galen, gave accounts of deformed bones and soft limbs in children as early as the first century AD. The doctor Soranus, also in the first century AD noted that certain kinds of pain were more frequent in Roman children than in Greek children and concluded that this was due to the poorer hygiene in Rome.

It was not until the seventeenth–eighteenth centuries that there were again accounts of soft bones in children, this time in England, where they suddenly appeared in an occurrence of almost epidemic proportions, often with fatal consequences for the sufferers. Soon the illness was known in many countries, where it seemed to follow in the wake of industrialisation. Children were shut up in factories during all the hours of daylight and were thus deprived of what was already only sporadic sunshine in the industrialised countries. For several hundred years thousands of children were crippled and died in appalling agony

Folk medicine and gifted local doctors gradually tracked down the inexplicable connection between sunlight and health. Children in the country, and especially in fishing ports, were not afflicted to the same degree. Cod-liver turns up in more and more places in medical annals, but it was not until 1919 that the English doctor, Edward Mellanby, conducted a definitive experimental investigation which laid the foundations for an effective prevention and treatment with the mysterious substance in cod-liver oil.

Even though the functions and significance of Vitamin D are still not absolutely crystal clear, there is no doubt that lack of sunlight reduces markedly the human being's survival mechanisms. Therefore it can also be concluded that change and adaptation of the human skin colours has happened very rapidly, seen from an evolutionary point of view. Our relatively slow population of the globe has occurred in time with our adaptation to new geographical circumstances.

When Indians, Chinese, Africans, Aboriginals and Scandinavians look the way they do, this is due therefore to a very precise adaptation both to one's native region and the original mode of existence. Much suggests that we have serious problems if we move around too much geographically, or change our life-style. And that is definitely food for thought in a globalised world.

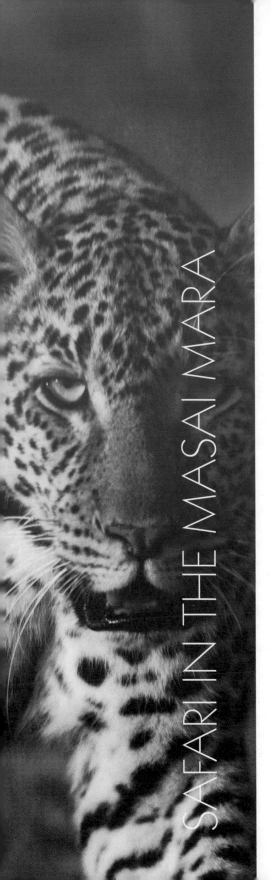

SAFARI IN THE MASAI MARA

For some life is a burden,
for others a gift,
for most a burden-filled gift.
The burdens are inherited,
the gifts you have to work for.

Awkward compromises

The sun is high in the sky as, sweaty and black and blue, we can finally observe the waves of heat vibrating over the sun-sweated plains of the Mara. The family dozes in the back of the battered car which is not a Land Rover but a Toyota. A safari vehicle is, by definition a Land Rover, but the Japanese have also, as in other areas, overtaken us where the asphalt ends.

The regular inhabitants of the open savannah have headed off to the thorny umbrella acacia and the doubtful shade of the sporadic desert date palms that invite one to a feast of luscious green grasses as payment for the daily fertiliser. The midday heat transforms the grazing zebra family into a riot of black and white zigzag stripes.

We are so close to the equator that the temperature on the plateau can sneak up to 40°C in the middle of the day. People and animals are wise enough to observe the necessary precautions. We drink and sweat. Fortunately we've brought a large cool box with delicious, refreshing drinks. Along the road a family of ostriches runs parallel with the car. The thick, insulated wings are raised up high so the wind can flow over the bare skin underneath and cool down the blood. Antelopes and gazelles turn their hindquarters towards the sun and avoid direct sunlight on their broad, dark flanks, and some are even lucky enough to find shade under acacias or candelabra trees. Thus, each of us, animals and humans has our strategy for maintaining the balance of liquids.

Even though there was no shortage of experience or renewed warnings about the condition of the roads, we had chosen the spectacular and quite impossible route over the Mau Escarpment where large stones and mud holes impose merciless demands on suspension, traction on all four wheels and two men on the bumper when things go completely pear-shaped. Here on the ridge where the clouds are thickening, there is no lack of water and most months of the year offer a bounty of sticky mud-holes.

All the roads in Kenya are quite atrocious, so you might just as well get something out of the trip. The alternative is the highway through the Great Rift Valley, where fragile minibuses fight a desperate battle with thousands of holes in the asphalt. The road to the Masai Mara seems to resist every incursion by road-building schemes, financial aid and private initiatives. The roads are only slowly improving, even though one often sees bulldozers, heaps of gravel, asphalting machines and the related fencing of newly laid roads. Perhaps it does all hang together, that a prominent politician with a number of important ministerial posts behind him just happens to own the airline ferrying tourists quickly and effectively to the Masai Mara. A business with no future in competition with cheap land transport on a fine asphalt road. Corruption continues to stick to Africa.

The children sit quietly in the back of the car, hardened by numerous safaris and many hundreds of hours of driving on African roads. They know all the animals, so I try to initiate them into something as boring as the significance of that substance in short supply: water. This undervalued fluid does of course run through the western world in unlimited quantities through millions of chrome taps, which is why, on first encounter, it seems inconceivable that the Masai word *nkai* means both God and water. No water, no life, it's as simple as that, and all the inhabitants of the savannah, wildlife as well as people, are daily concerned with the subject. Water occurs in measured doses and in annual fluctuation, to which everything living thing must adapt. If the fluctuations are too great or the rains are delayed, life starts to drain out of the plants, the wildlife and human beings.

A single glance out over the savannah bears witness to the significance of water. Small green oases are to be seen everywhere, where the water collects in hollows and where evaporation is retarded under shady trees. This is exactly where the wildlife stay during the hot midday hours. The wildlife must, on one hand, be insulated from the cold of night, and on the other, be able to escape the heat of the day without losing the precious water. There are numerous solutions to the problem even though sweat is out of the question in a dry environment like this. Every species has its own combination of solutions, which determine where the species in question finds itself most comfortable. Should the conditions change radically, the wildlife have to move to a suitable combination of conditions or die, but it is far from all species which possess a strategy that makes survival under extreme conditions possible.

I try to explain the colours of the game as a compromise between camouflage and cooling in the same way as the human being's skin colour is a compromise between protection from the sun and the formation of Vitamin D as discussed in the preceding section. The game must thus avoid both predators and water loss or overheating, even though the solutions each pull in opposite directions. The gazelles are therefore the most lightly coloured or white on the belly, the lower flanks and hind parts, while the rest of the body and head are camouflage brown. The light colours reflect light and heat when the gazelle is standing up. On cold days and at night they can lie down, hide the white parts and chew the cud, while they offset heat intake with camouflage. Theoretically a thick white pelt would be an ideal solution as protection from overheating and fluid loss, but white variants tend to end up in the stomachs of predators.

Nature is full of compromises, and no individual solution is ever absolutely perfect, as many, often mutually contradictory, conditions have to be simultaneously allowed for. The best solutions are always in mutual competition, and the running adaptations are never concluded, but create the dynamic evolutionary development.

My daughter asks the 'killer' question: "Okay, what about the zebras?" Black and white stripes, hot and cold, yin and yang? How is that supposed to fit the theory? Many explanations have been put forward, but most biologists are, however, agreed that the stripes offer good camouflage. In addition, complementary colours function as warning signs in nature. Humans and animals do not see alike. Every eye has developed in its own way, for different purposes. We do not see the same wavelengths as lions, owls and insects, only the wavelengths which evolution has found it worth concerning itself with. We are good at simulating the objective sight of a lion, but not the image which it forms in its brain. For this reason, we need to observe the conditions directly in nature. For instance, the Masai have stated that, for many animals, red is conceived of as frightening and produces an effect of terror. Therefore they have painted themselves with ochre for many hundreds of years. Today the ochre has, however, been replaced by thin red cloaks, which, moreover, also provide a certain insulation on cool days and nights.

Predators do not necessarily, therefore, see the zebra family as we do, and it has been conjectured that the stripes in a group of zebra run together in the predators' minds and prevent them from singling out a possible victim.

Entabeni Mountains offer fantastic walks and game drives.

No predator begins a hunt before the quarry is chosen. That would result in too many unsuccessful hunts and excessive energy loss. If a predator expends too much energy in a hunt it will die of hunger sooner or later. This is what happens with old or solitary animals which move towards an unobtrusive and often merciful death by starvation.

The common zebra – also known as Burchell's zebra – lives in many separate populations, which look markedly different. They do not only occur in the familiar, graphically pure black and white marking, but also in a variety of combinations with brown or light-brown stripes against the white. Each of the populations is, however, distinctive and differs clearly from each other in markings.

If one follows the herds of zebra from East Africa to South Africa, one can observe how the distinct black and white East African variant becomes more 'muddy' the farther south one travels. Every zebra has unique stripes, and in any case, every population has a unique 'uniform' because the individuals that stand out too much are simply picked off and eaten by predators. This supports the camouflage theory. There are clear limits to how much an individual can stand out, if the camouflage system is to function properly. The foals, however, have markedly divergent markings in their early life. They are browner and more visible, which probably puts them even more at risk. So what advantage is this for the zebra? Difficult to find a sensible answer.

Henceforth we can speculate on why the white heat-repelling stripes do not fill a greater area, and the black ones a correspondingly smaller one. As far as camouflage is concerned it should hardly be a problem. Theoretically it would be a better solution, but since it doesn't work that way we can, as faithful Darwinians, maintain that there is a circumstance which has not been fully elucidated, or that nature is not as smart as we are inclined to believe.

For all creatures it all comes down to keeping oneself alive long enough to put progeny into the world and ensure they have a good start. The genes must be sent further into the future. This also applies to securing adequate food and water, avoiding ending up as prey for others, and finally mating. It is these qualities that ought to be reinforced through evolution and natural selection. These characteristics are developed in a coordinated fashion, rationally and dynamically, while the energy is always used where it can do the most good. This means that many characteristics end up as something run-of-the-mill because more important things have been focused on in relation to the aim of survival and procreation: the only valid criterion for success.

Many of the animals of the savannah cannot satisfy the need for liquids through food and must daily seek out watering holes at considerable risk of being stolen up on by a predator. The giraffe is no exception but compared with the other wildlife it has an extra long way down and must drink in an awkward position which handicaps flight. It is seriously exposed to predators and must simultaneously struggle to keep the blood away from the brain while the head is held down. It must simply divide up its drinking into many sessions in order to avoid a cerebral haemorrhage. This is evolution's price for the long neck, which has given the giraffe a private, exclusive dining room high up in the trees, out of the range of any rivals. At first one may conclude that the giraffe's advantages in having an expanded food niche outweigh the difficulty with drinking and all the many physiological adaptations needed to pump blood up two to three metres above the height of its heart.

The giraffe has used a great deal of energy in developing its neck. Its entire physiology is appropriately adapted. One example of this is the intake of oxygen, which is rendered difficult because of its long passage. Theoretically it should demand that the giraffe's respiratory reflex should start the inhalation of fresh air, even before the old, oxygen-starved air has reached up through the pharynx. The situation is reminiscent of a badly out-of-breath runner who is hyperventilating to compensate for the inadequacy of oxygen. But this would be incorrect, since the giraffe takes in an extraordinary amount of the inhaled air's oxygen, so that it can comfortably await the completion of exhalation. The evolutionary explanation is that the individual haemoglobin molecule in the giraffe – that which, in the red corpuscles, takes up the oxygen – is smaller than in other animals, and therefore the total amount of haemoglobin in a giraffe has a greater collective area to which the oxygen can attach itself. One cubic metre of sand has an infinitely greater particle surface area than a cubic metre of fine shingle.

When one observes the giraffe, one easily gets the idea that evolution always gets the most out of the available possibilities. Thus, unlike the rhinoceros and the elephant, the giraffe has developed formidable eyesight, the optimal effect of which is due to its position on the top of a long neck. In addition the male giraffe uses its neck like a hammer handle in the battle over mating rights. Here evolution has also contributed with an extra refinement, at the very top, one might say, in that the giraffe's cranium grows and increases in thickness and weight throughout its life. Long neck and large head simply provide the best mating opportunities. The genes must be passed on.

We come a long way around the nooks and crannies of evolution, where logic frequently makes good sense, which, however, is not the same as scientific proof. Often we have to be content with making things probable, and frequently there are many probable choices available. We cannot separate characteristics from each other when we set them in evolutionary contexts. It is the sum of characteristics which is forced down from generation to generation – not just individual elements.

When, for instance, nature, summoning all its massive power, develops a complicated brain in order to control concrete physiological functions – such as a trunk – there is a certain logic in that this excellent organ is made available for other purposes. In this way a number of new possibilities offer themselves, which in turn unfold in different ways from species to species. This is not especially brilliant; it's just the way it is. Nevertheless for many centuries there has existed a belief that nature is a particularly brilliant mechanism where everything fits like a glove. This notion persists to this day and it casts a shadow over rational perception.

We have to go back to the brilliant Swedish student of nature, Carl von Linné (usually known in English as Linnaeus) who was born in 1707, a good hundred years before Darwin, to understand bowing before the greatness of nature. Linnaeus set about rendering intelligible God's magnificent work of creation through the systematisation of all animals, plants and minerals, which he named and classified in relation to each other. Everything could be located within the system which is still in use today in a modified version, and thus it was made clear to everyone how magnificently everything fitted together. Linnaeus's system also contains the seeds of the theory of evolution, because it illustrates clearly how closely the species are interrelated. One does not need to look too closely before catching sight of a gradual development, but it was to the maligned Frenchman Lamarck that it first occurred while he was excavating molluscs which changed markedly from one geological stratum to another.

Even though Linnaeus never wrote anything concerning this kind of heresy, we know that he was extraordinarily gifted, diligent, and at the same time a scientific researcher, who would not even shrink from making the human being akin to the apes, when the system showed that this was in fact the case. Fortunately by this time Sweden was 'reformed' and so he had no cause to worry about being anathematised with a papal bull. It is hard to imagine that the notion of the development of species hadn't crossed his mind: the idea must have given him sleepless nights, just as it did his colleague in England a hundred years later. I don't know if I've managed to explain this sufficiently, since we are all born with the idea of simple, fantastic Nature, in which everything fits together, but the deeper we thrust the spade into the ground, the faster simplicity disappears. In order to understand the totality we start with the detail. But it is so incredibly difficult to tear oneself loose from … from the detail!

And Nkai said, "Let there be water!"

After a long, hot day, we finally come in sight of the safari camp by the River Mara with the Oloololo Escarpment in the background. The ridge is my permanent navigational reference point, which looms up quite anonymously at the horizon and reveals the location with an accuracy of a few hundred metres.

David is a Masai and manager of our first permanent safari camp in the Masai Mara. A gifted young man with the Masai's typical athletic exterior, he has lined up the personnel to receive the Big White Boss from Denmark. The sun continues to shine from a cloudless sky, we are tired and hardly have the strength to go through the lengthy ceremony of welcome. I hesitate a shade and flop back into the worn seat of the old Toyota, as the grunting of the hippopotami bids me welcome to the real Africa; the Africa which should remind one of the time before the world went out of joint if indeed that is the case.

The personnel consists predominantly of new people, and they are all dying of curiosity to catch a glimpse of their boss's boss. In Africa a boss is something really big, something best demonstrated through ascetic abstinence concerning every form of physical labour, a huge desk with metre-high stacks of paper as well as a dark suit and tie. A white chief is even grander, and if into the bargain he has a residence overseas, we're approaching presidential status, a humble origin in Danish soil notwithstanding.

There's hardly anything presidential about my appearance in worn-out sandals, worn-out shorts from Sportsmaster and a locally made khaki shirt. There are a lot of sideways glances, but one is reluctant to appear discourteous. I greet everybody politely and warmly, ask about the job, chat a bit, fire off a couple of jokes. Everybody laughs a lot, possibly out of duty; perhaps because

Africans laugh easily. I sense a certain reserve. My face is examined thoroughly for the unmistakable symbols of the worthiness of age, such as wrinkles, grey hair, missing teeth and other forms of decline. All of these are honourable decorations in African eyes. Africans have just as much difficulty reading our faces and determining our age as we have when it comes to Africans and other foreign ethnic groups. I am still lacking these prominent badges of honour, which I, moreover, have made a certain effort to conceal, and leave behind therefore a not insubstantial perplexity in this milieu, where everyone has a set place in the age-determined hierarchy.

On the savannah one bows the head out of respect for the old. Even highly educated people return to the family and clan and live on in the staunchly conservative tradition to achieve the unavoidable respect due to age, which is unrivalled by all the gold or wisdom culled from an outside world. Age means everything. Hundred-, maybe thousand-year-old myths, rituals and traditions maintain that shrewdness and experience that come with the greying of the hair. Here people still stand up for the elderly on buses. So the fact that I am middle-aged or rather quite old by the local yardstick, has probably been a great relief, for who could serve a youth?

David loves to exchange observations on traditions and rituals, but my abilities often fail to pass muster since what I have to offer is simply too alien or too foolish for comparative analysis. How are you supposed to explain the philosophical significance of the Danish clap-hat to a cattle-herding nomad?

How do you explain to him, in particular, that we have definitively disposed of age? We have no need of the elderly, and certainly none of the truly old whom we would rather isolate from the family as long as possible. There is no prestige in old age; technology has defeated experience and we fight a hard, resource-consuming battle against the unmistakable signs of age. Whether or not we want to wear low-waisted, figure-sewn T-shirts or stone-washed jeans as an eighty-year-old, we simply do not wish to grow old: "Hey look at me, I'm old enough to sit and loaf around under a shady acacia."

David shakes his head and nods profoundly, but hasn't the slightest idea what I'm talking about. It's just too far out. I throw myself into graphic descriptions of how we cripple our bodies: about how we willingly submit to having ourselves cut in the face, breast and stomach to straighten out the skin, colour the hair, drag ourselves through numerous hours, sweating like pigs, in fitness centres, run ultra-marathons with our tongues hanging out, guzzle slimming powders to make room for that extra burger, while a global pharmaceutical industry spends billions to develop medication which removes the signs of age as gently as possible. Here we are apparently on some trip which is insanely funny, but quite untrustworthy. I really am a sly dog; he himself would never dream of making jokes about such matters.

We talk a great deal about our respective worlds, not least about the unheard-of controversial topics such as equality, circumcision and homosexuality, on which David represents opinions which in Sweden would have landed him in psychopathic custody for life.

Time after time, we, or more correctly I, return to the question of the position of the elderly, which is naturally synonymous with elderly men. Where the elderly enjoy massive prestige, it all hangs together with the Masai's conservative life strategy. They have no need for development, new kinds of thinking or enterprising young men who revolutionise the world. Alright, they would like to breed a cow which gives more milk, but it certainly isn't on the cards that nature can or should be controlled. It's a question of understanding nature and using it to the best advantage in the prevailing conditions.

For several thousand years they and their forefathers have lived in the arid savannah regions and semi-deserts where experience alone has ensured survival. They have lived with the threat of natural catastrophe, which is deeply embedded in religion and behaviour patterns. Drought and failed water supply are a latent threat, and this applies particularly to the great ultimate drought catastrophe where even bulls and people die. They know that it strikes at intervals of approximately 100 years, and here one has the need of the living hard disk: the elderly who have experienced a number of fairly major catastrophes, and can remember in detail accounts from the grandparents of the last 'bull death', as they call the greatest drought. Experience alone can defeat the catastrophe and therefore experience means everything. Water, survival and experience are inseparable concepts.

The combination of experience from other actual droughts and historical knowledge of patterns of action from earlier catastrophes ensure survival. This strategy has been successful. The people have survived through an established cultural pattern, which still maintains that water is God. Nkai.

Throughout all ages, water has played a key role in the life of the human being, consciousness and religion, right from the time when we wandered alongside the rivers, settled beside them and began to cultivate the soil. Outside the ancient city of Petra one can see broad, dry riverbeds, the city's original life-nerves. The water has periodically flowed so powerfully that the redoubtable ancient people, long before our reckoning of time, had to dig drainage canals to prevent themselves being flooded once the rain set in. It is part of the Bible's Promised Land, the 'Fertile Crescent' in the Middle East which flowed with milk and honey. Known as the 'cradle of civilisation', to which our African forefathers wandered, learned to till the earth, tame cattle and lay up the surplus for the bad times.

Along the waterways one finds a number of large, square sandstone colossi, which ancient people laboriously carved for the benefit of their supreme deity

Dushrat, a Middle Eastern pendant to Zeus. They are the abodes of the gods which are called *sabrii* in Arabic, which actually means water jar. Naturally I could not resist the temptation to tell the Masai about Dushrat and the water jars, which were the dwellings of the gods. All of which was immediately understandable to a Masai. There is nothing more important than water, and a jar for keeping that which is the most important of all is naturally a worthy abode for a god.

The word can suggest there may be something inside, but the fact that a sabrii is solid is of no particular significance, for a god is immaterial and can make his abode in anything whatsoever.

Water has been just as essential for our civilised forefathers in the Levant as to the Masai of our time. They have prized the water which has been the precondition for cultivation of the earth and thus civilisation. Without water, no civilisation. The entire Middle East is dotted with menhirs, obelisks, stupas or mazzeboth, as the individual stones are called, which all bear witness to our forefathers' worship of gods. From here they have borne the tradition with them out into the world, raised monuments and abodes of the gods which were intended to ensure water, harvest and prosperity. Since then the god has flown into the sky and earthly dwellings have been replaced with houses where the human being can make agreements with God. We call them temples, churches and mosques, but with the Masai it is still the same water god. But, then, are there others?

Circumcision and manliness in the sign of the lion

David and his Masai brothers' red cloaks provide a certain security, but even though the red colour is supposed to frighten potentially dangerous animals, the grass is no shorter for that reason. The combination of long grass, sandals and wild animals bothers me increasingly as the years pass. It makes me wonder how they trudge quite unworried over the savannah without taking seriously all those things which might lie hidden. Perhaps one becomes a coward with age, but I choose now to interpret it as great wisdom, and I wouldn't reject the idea of having a solid powerful hunting rifle to hand. Okay, it's a question of not putting oneself into a situation where one needs a firearm, a statement with which I have frankly diverted a good many people. Of course it's true, guns give a false sense of security, but even so!

Should one be unfortunate enough to cross paths with an ill-tempered elephant or buffalo, one should hang one's red cloak on a branch and flee, David explains. And what if one is wearing a boring, greyish-brown khaki shirt? Despite many years of experience, one cannot, as a city dweller, ever quite get used to the utterly fearless relationship with nature, which may not be a bad thing. As new arrivals we can learn to 'pretend' but we simply cannot sense

A Masai *enkang* (enclosure or village).

nature's signals fast enough and will always be fearful. Perfect fearlessness is only to be found in psychopaths and people who have been brought up as warriors from birth. Most Africans resemble us perfectly on this point. An African city-dweller can be scared out of his wits by a peaceful chameleon, simply because, throughout their childhood they have been threatened with the big chameleon – a local version of the bogeyman. The situation is completely different with those who live on, and from, the savannah. They are not courageous in the conventional sense, but have quite simply jettisoned fear through many years of cultural adaptation. An ability which also makes them among the best warriors in the world.

The Masai and other nomad peoples drive their cattle and exploit the savannah's grazing options in competition with the wildlife while they protect themselves and the cattle from predators and other dangers. Hunters and gatherers such as the Pokot in Kenya and the Bushmen of the Kalahari have refined the technique and they have a formidable knowledge of everything that occurs in their immediate environment. These people possess a quite exceptional ability to intercept at lightning speed all signals and any divergence from the normal state of affairs, and to react just as fast. In their surroundings they are trouble-shooters. Time after time it astonishes me how they can point at an elephant that I cannot see. It *should*, however, be possible to catch sight of an elephant.

When the smell of fresh acacia sap hits the nostrils one knows that there

Young Masai with circumcision décor.

are elephants in the thicket, grazing on the trees. If the wind is in the right direction the smell hits one before the sound. One can get to recognise the smell and many other signals, but not everybody can learn to handle the complex mass of information hitting all the senses simultaneously. This extreme ability to read nature recalls the capability we usually call intelligence, which is surely today regarded as a universal, inheritable quality.

The inhabitants of the savannah have simply adapted to their environment to a degree in no way inferior to that of the wildlife and only when one begins to examine these more closely, is it striking how it all hangs together, how behaviour, tradition and belief interact like interconnected water jars.

Then I see the lions. First the

waving tail, like the top of a desiccated leonitis flower, then the golden-brown back. This is far from the first time I have been close to lions on foot, but even so the adrenalin is coursing from the very ends of my hair to the big toes and back again. David is completely unaffected, acting as though he hasn't noticed anything and proceeding without turning a hair. As phlegmatically as possible, I say something on the lines of, "You know what? I reckon I'll just get a couple of pictures of those lions, but they're miles away!" The suggestion that I want to get closer occasions a collective grin in the little troop, who think we should not frighten the poor animals. There is a certain part of my anatomy I'd rather they didn't take (?)

Lions never attack Masai and if we move closer, they will probably get to their feet and plod into the denser scrub. Lions are intelligent animals that keep learning all through their lives and are capable of passing on their experiences to the next generation. Lions are like people, elephants and monkeys, completely defenceless as newborns and have to undergo a lengthy learning process before they can fend for themselves. Even young adult lions can look forward to a harsh existence if they are not part of an experienced family.

The lions in Masai country, which stretches across the savannahs of East Africa, have learned through numerous generations that it is best to stay a long way from the people of the savannah. The Masai have killed big male lions as an obligatory test of manhood for many hundreds of years, and it is just as certain as the continuing utterance of amen in church, that any attack on Masai cattle will be followed up with a lion hunt which will just as certainly end with the death of one or more of the lions involved. Here lions only grow old if they know that the human being is a more dangerous enemy than the hyena, but, on the other hand, not a rival. So: much less reason for confrontation. I am quite sure that lions are clever enough to tell the difference between me and a Masai. That won't be put to the test.

I tease my red friends that lions simply don't like their scent, but the curious thing is that even though they live in houses made of cow dung and have the most extremely limited contact with water, there seems to be no special body odour.

Slightly pressured, a couple of the young warriors get closer to the lions, so the great photographer can get a decent picture. While they are posing, two lionesses get to their feet and look at us. The young men turn and move a slight distance from the lions. One lion executes a little jump and ends up lying down in the shade of a croton bush, five or six metres away, while the other lies down on the spot. Yes, there it is, mutual respect.

I am absolutely clear that an outsider, a foreigner, can never get close, mentally, to a Masai, whatever that means. The bonds of family and friendship have been completely laid down by tradition, and standing room is not provided for the uninvited. These family ties are not like ours. A manly Masai's 'next of kin' are the men with whom he was circumcised, and with whom he has subsequently shared the path of the warrior. The relationship with one's own age group lasts one's whole lifetime. Every woman is the nucleus of a family, consisting of herself and the children up to puberty, together with 200 head of cattle, for which she and the children have responsibility and must milk. A man tends to pass most of his time on his own or spending the day with other men, but he chooses every evening which family shall be honoured by a visit, or, rather, with which woman he will share a bed for the night.

All in all, the family pattern reminds one a great deal of corresponding patterns in the surrounding wildlife, where the females are the backbone of the family and the males live on the periphery of the female units. This is the case among the lions, which play a major role in Masai mythology. The young male lions have to leave the mother group at a point when they are in the very first flush of youth, and must expect some hard years as lonely excluded warriors, before the smartest, best and strongest subjugates a group of females that can provide and care properly for him.

I sneak in on the controversial subjects of the position of women by taking the long way round. Answers and explanations are always presented without any beating about the bush. It is my European view which initially blocks what is absolutely obvious: that the Masai have nothing to hide. There is never anything furtive in their tradition: everything has an explanation, a cause. What is the problem other than European narrow-mindedness?

The topic for which I have been fishing for a long time, is that of female circumcision. Is there any meaning to that madness, or is it merely an old tradition which has survived in this conservative, male-dominated world, without anyone having the slightest idea why? There may be hygienic advantages associated with male circumcision, but that is not the case with the female operation, which is performed here in its most radical manifestation, where both the clitoris and the labia are removed. The relatively innocent circumcision of the men is surrounded with enormous seriousness and respect. For the rest of his life the circumcised warrior will be able to feel self-respect for his brave demeanour during the ritual, while the girls' violent and life-threatening sufferings are simply a minor issue, signifying merely that she is eligible for marriage.

Under cover of the adrenalin rush from the lions I pull myself together and ask the question absolutely crystal-clearly and without circumlocution. I ascertain, with some surprise, that there is certainly no hitch anywhere; the girls should not in fact wander around and develop lusts. Circumcision is merely a simple means of avoiding passion. Lust and love fit naturally together, and if one limits lust, it is easier to marry the girls to older men without too much trouble. The girls get a man who can support them, but they must take it in turns to prepare food for him and close their eyes for five minutes once in a while. My last interpretation gives rise to a lot of amusement, not to mention giggling, and it is suddenly clear to me that, – fortunately – things do go on behind the scenes, and that the young men and women can actually influence the process and have certain prenuptial contact. This is, on the other hand, not something that is talked about.

The calendar year is of no great significance to the Masai, who instead calculate in periods of seven years on average – adapted to the moods of nature. The oldest determine the length of the periods following long, thorough negotiations and considerations of how the warriors have acquitted themselves, and when there is need of a new, fresh generation of warriors. For this reason the girls and boys are circumcised in year groups, roughly every seven years. The age interval within the 'year' may, of course, be very large. An unlucky 13-year-old who was found to be immature in the first instance is thus not officially adult until he or she is 20–21 years old.

The girls are married immediately after circumcision. The young men have to wait until their term as warriors has passed, at the next circumcision ceremony. Marriage often also depends on whether his herd of cattle is large enough to support a family. His first wife will therefore always be from the next circumcision or maybe even later. The adult warriors then become 'elders' and can further work up the number of cattle with a second wife in mind. There is no upper limit on the number of wives, but according to Masai arithmetic there should be around 200 head of cattle per family group (women with children) since 80 per cent of the cattle die in the ultimate drought catastrophe, and the family's daily maintenance demands a minimum of 40 head of cattle, or the equivalent in sheep and goats.

Young love or not, the system results in by far the most women ending up with a somewhat older husband and many with one who is considerably so. This, however, fits with the picture that there is a certain tradition that old men's young wives can receive visits from younger men, but neither is that something people talk about very much. In return the husbands pay a 'dowry' to the bride's family of seven to nine head of cattle as well as rugs and tobacco. Besides this, meat must be made available for the men's feast: *olpul*. And yes, that's its actual name.

There are, therefore, considerable demands on the men and their success. They must have been warriors with a respectable reputation, have experience of life and wealth, the latter being acquired by their success in building up and defending a very large number of cattle, which can support a whole family, after the bride-price is paid. It is, therefore, the lot of far from every man to start a family.

We talk while we walk. The rapid tempo of the Masai suits me well; we only stop when I stop to observe a bird, a termite colony or big game. I have ceased expressing excitement. It means nothing to them. Their approach to nature is rational and devoid of romantic notions. They know all they need to know, especially danger signals, but otherwise they are blank in relation to the knowledge we normally find interesting. I would like to offer something in return for the family histories, and contribute with something or other. I knead the brain until I see about twenty weaver bird nests hanging from a palm. There is lively traffic around the nests where both males and females are busy raising the new generation. It's the industrious males who build the nests, I say with a marked emphasis on 'industrious males'. The reference to the lazy Masai, who have the women build their dwellings, is picked up and causes a certain merriment. Several of the nests are unfinished and abandoned which gives me occasion to explain how the females evaluate and prefer the good nest-builders among the males, which otherwise have to start from scratch again and again until the business is learned properly.

The family patterns of the Masai, as I have previously remarked, are not so removed from those of the animals in the surrounding environment, where the gerontophile tendencies seem to be in the best of health. This does not apply merely to the lions, but also to the giraffes and the elephants, where, to an overwhelming extent, the females mate with elderly partners.

The life cycle of the elephant greatly resembles that of the human being, but the males should really be around 35 before the females will mate with them. There are exceptions, as there always are in nature, but, overall, research has shown a marked preference for age. The elephant has to have demonstrated its survival skills before it gets a share of the spoils. This state of affairs is actually supported by genetics, which arrange things so that growth continues throughout life. The largest male tends also to be the oldest. If a female prefers the largest elephant, she tends to get the oldest.

A corresponding situation occurs with the giraffe, which, as mentioned previously, is equipped with an especially refined and genetically conditioned age advantage, in that the bones of the skull grow thicker and heavier throughout life. The heavy head on the top of the neck is used as a sledgehammer against other males in the struggle for mating rights. The oldest head is the largest and hits the hardest.

These kinds of stories always get a laugh. When animals actually demonstrate some kind of 'cleverness', which is not inferior to good Masai traditions, then laughter with a certain self-satisfaction is possible. Yeah, right, clever animals, no doubt about that. When the amusement subsides, one observes a general tendency for the mating rights to follow experience in species with large brains and a long childhood, in which much has to be learned. This is true, for instance of elephants, many members of the cat family, most apes and whales.

When one has, oneself, reached a certain age, it can be tempting to argue for the advantages of the prior claim of older men, but I would be content to fasten upon the culturally determined. It is of course a well-known fact that the tendency exists in a good many cultures, both historical and more modern, but (fortunately), I can safely leave the argumentation in the hands of the anthropologists.

When the fragments are pieced together, the continuity in the Masai tradition gradually becomes apparent with exceptional clarity. Rational and coherent, it is absolutely what we usually call culture, not least in the light of our own, western culture's lack of cohesion. Here there is a direct link between productive forces and superstructure as old Marx would surely have analysed it, and so I shall refrain from discussing which is the product of what.

The supporting pillars of culture are experience and skills. One must demonstrate one's competence, his or her ability to survive and provide for his or her family, as in so many other places in the natural world.

The Masai are a tiny ethnic group with enormous success, if measured in yield by area – kilos of human flesh per square kilometre, to express it in a manner totally devoid of feeling. Their culture and mode of existence are adapted to Africa's marginal terrain: the savannah, those dry and semi-arid regions that are not suitable for agriculture. Here they have learned to adapt and optimise to the utmost. They know the humours of nature, not least the drought, from the frequently recurrent kind when the calves die, to the rare catastrophe where humans too perish.

They have developed a comprehensive and effective strategy which puts them in a position to deal with most situations, and to be resigned when they have to yield to nature's capriciousness. Their culture functions the same way as the evolutionary selection of the best suited.

Through the recent millennia, the Masai and the other Maa-speaking people have spread across the East African savannah. There is still considerable mystery as to their origin, even if one supposes that they have a cultural point of departure in Ethiopia, whence they followed a route through the massive, dry East African Rift Valley in present-day East Africa. The interesting thing is that Maa races probably constitute a small independent enclave in the middle of a hostile sea of crop-growing Bantu with an entirely different culture and tradition, which seems utterly incompatible with nomadism. How has a vanishing minority managed to persist in the face of an overwhelming majority?

Their culture is resilient and much suggests that it has been improved and optimised to perfection en route. It has developed from a blend of cattle-driving and gatherer culture, possibly with a little agriculture, to pure cattle nomadism, which, in the final stage has been refined with a simple diet, division of labour and warrior culture.

Division of labour has shifted the entire maintenance of the family onto the shoulders of the woman and high-level politics to the man, who concentrates on the art of war and leadership. The austere diet consisting of milk and blood supplemented by a small amount of meat on special occasions represents a complete minimising of time, or rather energy, which goes into the provision of food: a minimising of the job of searching for food which can be compared to that of the ruminants. From childhood, the men have concentrated on the art of war and have been brought up to a state of absolute fearlessness, extreme consciousness of themselves and their own rights, which, naturally come before those of everybody else.

The Masai were once so feared that the great journeys of exploration in the nineteenth century, such as those undertaken by Speke, Burton, Grant and Livingstone made wide detours to avoid them. The highly numerous Bantu tribes didn't stand a chance and had to endure the cattle nomads making increasing inroads into their agricultural land. Why should one be content with the dry semi-desert when there were lush green fields available? There are examples of wars and disputes during the Masai incursions, but the Masai reputation was so terrifying that actual fighting was seldom necessary.

Everything interlocks in this culture, simple and, as it were, cast in one piece, which doesn't contain elements of development or constant change. The framework is fixed and the way the problems present themselves is resolved within this. The question is, what happens when the framework is challenged? Are they capable of finding counters to modern, flat-screen civilisation? Can they, like the lion, adjust to another kind of prey in hard times, or must they perish like the cheetah, which is lost without its habitat. Could it be that the one-way culture has laid its genetic trail, or is the culture a result of an ordinary, average 2014 brain, which has remained unchanged through a long period of time?

In Karen Blixen Camp we employ Masai women who have been removed and sent to school before circumcision. NGO organisations remove – with the government's approval – the most gifted girls from the local schools and take them into a secure institution. In their turn the Masai try to have the brightest girls circumcised and married as quickly as possible to avoid unpleasantness and protests.

The times they are a-changin' and it has become more widespread to use a gentler ritual circumcision where a small quantity of blood is sufficient to legitimise the condition of membership, whereas those who remain uncircumcised pay the high price of rejection.

Nature is a tinkerer

The sky is far too blue. We are waiting for rain. So are the fifty hippopotami in the small hollow. We regard them as our hippos, because they just happen to spend most of their lives here, where the River Mara flows past Karen Blixen Camp. They don't have any real influence on their own situation. They were born here and will also die here. The adult males can migrate downriver, but space is limited after a number of years with masses to eat.

One can see clearly the contours of the hip bones, which stick up like small bumps where the hindquarters are normally round and beautifully fat. We scout for clouds and discuss whether it is possible for us to get food for them. This is a Labour of Sisyphus, harder than rolling a large stone up the Oloololo Escarpment in front of us, We have simply too many of them. Fifty hippos, each of which require forty kilos, or, in total, two tons of fresh, highly nutritious grass per day. And where are we supposed to get that from?

The females have adjusted the breeding period according to the rains, or rather the rain which should have come, and all the small new hippos have arrived. We count ten of the little guys on the opposite bank. This is not promising. Soon there will be no milk for them. We have already dragged two heavy half-grown corpses away, so the smell will not annoy the patrons. Hippopotami succumb quickly if conditions are unsuitable.

The hippo has a stomach in three sections, resembling that of the ruminants. It can therefore be content with foraging for relatively few hours at night and relaxing for most of the day, while the food is digested at the animal's leisure. A splendid way of arranging things: a few hours work and plenty of time to take it easy. Where Stauning (the legendary Danish statesman of the 1930s and '40s who became the first Social Democrat prime minister) demanded eight hours of work, eight hours of time off and eight hours of rest, the hippo has been a pioneer and has arranged matters so it has five hours work, and nineteen hours free time and relaxation, since it doesn't differentiate too much between the elements of relaxation in the agreement.

Hippos and ruminants extract the maximum energy from the food. They do not therefore need to eat that much and thus minimise the energy spent on foraging. Just like the Masai in fact. When the dry season sets in, there are, however, severe problems with the strategy. Where the elephant and zebra can increase the quantity of food consumed, the hippo and the ruminants have to content themselves with filling up the stomach and waiting till the digestion is complete. When the stomach is filled with dry grass, poor in nutrition, the results are rapidly apparent, the animals starve and lose weight.

Everyday we cast an anxious eye on the hippopotami. They look peaceful and idyllic lying there basking in the muddy water with their heads resting on each other. Their voluminous bodies prompt thoughts of their closest relatives

In most cases the game will pay very little attention to the game viewers.
Inset: A luxury safari lodge.

in the world's oceans. Whales share ancestors with the hippos, in the same way that we share ancestors with chimpanzees and gorillas. If we go a little farther back we all share an ancestor among small shrew-like mammals. The mammals got more room on the planet when the dinosaurs suddenly died out 65 million years ago. A mass of habitats became vacant and the mammals had a busy time moving in. Mammals represented evolution's most recent solution to the problem of parental care: the eggs were fully developed inside the mother, and the young could then be carried on the back or in a pouch in the stomach. The new species could bring their progeny to food and protect them at the same time. Birds and reptiles, on the other hand, are, as egg-layers, forced to stay within proximity of the nest, without, however, being able to protect the eggs or young effectively from predators.

The mammals developed over the many succeeding millions of years and adapted to the various ecological conditions. The seas rose and fell in step with the climate, and for long periods large areas of the present landmasses were covered with seas, bogs, lakes and marsh. During these periods the mammals adapted themselves to a life near, in and with water. Everything considered, the closest ancestor to the hippopotamus and the whales was an amphibian which diverged from the ancestors common to pigs, camels and savannah antelopes. One line of development continued its search for food on land, and became the hippopotamus as we it know today, while another line opted for searching for food in the water and converted completely to the sea. Traces

61

of the earliest whales are around 35 million years old, but both whales and hippos have since undergone many stages of development to reach the present life forms.

Where the small West African pygmy hippopotamus is the only close relative to the common hippo, the whales have many, as, during the same period, they developed numerous families. It is often the case that each whale species has complicated family patterns with uncles and aunts, and at the same time an extensive communication system. Most of the families have developed large brains, which can solve complex problems and put them in the position to change their own circumstances and possibly update strategies to counter changes in their environment.

This time the rain came almost as foreseen and only the weakest perished. The problem developed because the period leading up to the normal rain was unusually dry and hot. On one hand the hippopotamus has adapted to fit its niche like a hand in a glove, but on the other hand the niche is so narrow that there is no room for flexibility. The hippo's home is a stretch of a water course, a water hole or the shore of a lake. Here it spends all the hours of daylight and about half the night. The remaining time it grazes as near the water as possible, even if this means moving in a radius of five kilometres from the water. In favourable times with sufficient rain the hippos spread out into all the vacant habitats, which gradually become over full and the grass is cropped down to the last stalk.

The small regular variations in climate will, as a rule, be sufficient to keep the population under control. Sometimes, however, the population becomes really vast and so it's merely a question of a few weeks without rain before things go seriously wrong. There is nowhere left to go; everywhere is occupied. When catastrophe strikes they go down like ninepins, one after another, starting with the young and frail, then the females and finally the biggest males. Exactly the same as with the Masai's calves, cows and bulls. The hippo has occupied a very specific niche in nature which has no alternatives, and therefore it is hardly equipped with any special talent for thinking. What would it use a flexible, creative brain for? The Masai, elephants and whales have alternatives and therefore also have a use for a brain that can differentiate between them.

Nature couldn't care less how many hippopotami survive, or even whether the species survives at all. If the changes in conditions are successive, the hippo might be able to keep pace with the development, or else there will be a vacant niche for other creatures. Exactly the same conditions in which the hippo took over the niche in its turn.

In this connection the brain can also be regarded as an organ without any special value, compared to all the other organs. Some species are blessed with a complicated brain because they can use it for a variety of problem-solving possibilities, which nature makes available. Others have lined up all the 'solutions' beforehand and can spare the brain's very high energy use. If the brain cannot increase the survival of the species, it is superfluous, but this does not, however, prevent us humans from attributing to it a superior overall value. If this notion of the brain is correct, does it also apply to other species which use it in a corresponding fashion?

The immediate impression tells a story of the hippo's formidable powers of adaptation which is consistent with the idea of the Greatness of Nature. When the time horizon is made a factor in the calculation, it is possible to see the inescapable progress towards disaster, which, most of all resembles a completely unnecessary waste of resources. This picture is repeated time after time when one gives it a thorough inspection. No adaptation is perfect, with or without a brain.

Think if the enthusiasm we express when we see examples of certain organisms' fantastic camouflage was really justified. Think if a stick insect in a tree, a moth on a tree trunk, a snake on the forest floor or a penguin with its white belly on the sea surface was invisible to its potential predator. Nothing, however, is invisible or perfect, but in spite of this one should observe closely. That is precisely the whole point.

The scholar of nature, Linnaeus confirmed the Greatness of Nature and God's work of creation through meticulous observation, which, step by step, created order in the objective world and made it understandable to ordinary people. Many have followed in his footsteps, and we tend to call his procedure science. The weak point is that people find it very easy to confirm their own suppositions and forget to look more closely for flaws in the pattern. If everything in nature fitted together properly every life process would come to a grinding halt.

A lion has difficulty bringing down a strong healthy quarry, which, therefore, has a fair chance of surviving and bringing a strong healthy progeny into the world. The next generation of prey make life even harder for the lion, who, on the other hand concentrates on easier prey: the wounded, the weak and the starved individuals. If the lion can't manage the task, it must succumb. Other lions do manage the task and have vigorous offspring, capable of surviving, which make life difficult for their prey. This is how the dynamic of nature functions. Mismatches create this dynamic. It is never absolutely perfect and just as well.

By far the majority of the young perish before they manage to breed. An enormous waste, which clearly bears witness to the fact that the adaptations are far from optimal. Half of the young hippos leave this life before they reach adulthood, and by far the most of the gnu calves end up in the bellies of predators. When the termites open the colony and send new generations out immediately before the rainy season, great flocks of birds, mammals and sometimes people gather around the fluttering insects to eat 99 per cent of them. Nature and adaptations should be regarded as dynamic, in eternal movement.

If we look a little more closely at the hippo's behaviour, we may perceive new factors which can contribute to our understanding of its adaptation and lack of the same. The bull is much larger and heavier than the cow, but its hind legs cannot support its enormous weight during mating. The problem is solved by moving the mating out into the water, where buoyancy provides the solution. Here a new problem emerges, because the enormous weight forces the cow under the surface. During the entire sexual act the cow fights a desperate battle to avoid drowning.

Curiously enough, the hippo's adaptation to a life in the water is not absolutely perfect. The animal cannot swim. It can exploit the buoyancy and splash about a little, but actual directional swimming on the surface is something of which it is incapable. Sudden changes in the current and unexpected waves caused by heavy rains are potentially fatal. If the hippo but once ends up in a powerful current it is lost and will drown. This happens at regular intervals, but not enough to endanger the species.

The bare skin has various adaptations to water, in which it must submerge for much of the time to avoid sunburn. But during the drought, when it needs an extra meal or it has to go farther afield to seek out food rich in nutrition, it risks serious sunburn, and if its habitat dries out so severely that full submersion is no longer possible, then it will die of burns and the consequences thereof. Drought scenarios are far from unique but the hippopotamus as a species has probably most to gain by being like it is, with the existing advantages and deficiencies.

Many elements are thus not so smart and well-adjusted as one might first think. When evolution has not solved these problems it is because they are minimal in relation to the overall survival. To put it another way, the hippopotamus has, in this way, more to gain than to lose in developing thin skin, great weight and small legs together with a relaxed existence with a low pulse and low metabolic rate. Evolution pushes the hippo in the direction in which it gains maximum value for energy expended. But then you can't have everything.

The French biologist and Nobel Prize winner François Jacob is a clever man who writes beautifully. His little book about evolution has been on my bedside table ever since I heard him in Copenhagen in the mid-1980s. In this book, in a somewhat unimpressed tone, he calls nature a *bricoleur* or tinkerer, and makes a break with the old conception of 'smart' nature, which, in the course of years, has been hawked through a plethora of books in colour and nature films. This point of view corresponds somewhat to my description of nature's dynamic and adaptation, which is often unique but never perfect. Jacob however, sees nature as a rag-and-bone business, a *bricoleur*, who, in contrast to an engineer, is a kind of dexterous part-time dabbler who collects every conceivable kind of junk and subsequently puts it all together in unsalable ways.

In the English edition of his book he writes: "In contrast to the engineer, evolution does not produce innovations from scratch. It works on what already exists, either transforming a system to give it a new function or combining several systems to produce a more complex one. While the engineer's work relies on his having the raw materials and the tools that exactly fit his project, the tinkerer manages with odds and ends. Often without even knowing what he is going to produce, he uses whatever he finds around him, old cardboard, pieces of string, fragments of wood or metal, to make some kind of workable object." (*The Possible and the Actual*, p. 34 Pantheon Books, 1982.)

The development and adaptation of organisms is controlled by instances of chance and accessible materials. Evolution extends that which exists, which is pressed into the desired form, that is far from optimal or refined, but is, thus, usable. As Jacob says elsewhere, "To manufacture a lung with the aid of a piece of stomach tissue is exactly like making a dress from Grandmother's curtain."

Four hundred million years ago, during a period referred to by geologists as the Devonian, some pretty curious fish were swimming around in the world's oceans. Some four to five times the bones in their fins divided in two, so that the fin came to resemble a lobe. Naturally we have given the fish the name lobe-finned. At one point, life on land was so attractive that the fish began to make incursions onto terra firma. There seems no doubt this occurred in much the same manner as with the small African mud skipper which hops around in the mangrove swamps on its front limbs. The fish have become the dry land's vertebrates – reptiles, amphibians, birds and mammals. The lobes became arms and hands with humans, the front limbs of many animals, birds' wings, and, in some cases they disappeared, as, for instance with snakes.

Nature has recycled, reshaped and optimised the existent possibilities which, in these instances, we call homologous characteristics. But in by far the majority of cases, the development is not as visible, precisely because the stomach is made out of a curtain. When one lifts it from off the individual, the *bricoleur*'s work becomes even more apparent, since it can be found everywhere, just as long as one observes, just as with the algae and the fungus that combined as lichen, not out of love but of necessity.

Like millions of other people I am a lover of Puccini's *Nessun Dorma* which you can find on my shelves in various versions, by Italian and Spanish lyric tenors as well as by an English reality star. I shall not pretend to know why people experience a pleasant frisson when sounds are combined in a particular way, but one of my *Nessun Dorma* 'variations' stands out from all the rest. It is arranged for the ancient instrument of the Australian Aborigines, the didgeridoo, and it sounds incredible, *bricoleur* or not. The didgeridoo is a small

Migrating wildebeest (gnus), a truly spectacular sight.

hollow eucalyptus trunk which the Australian indigenous people need only to gather from nature and blow into. The tree is called the mallee, but this term covers a much larger group of trees which are adapted to Australia's arid and nutrition-poor desert regions.

Shortly after the tree has developed an actual trunk, the termites begin to gnaw at it from inside, and gradually as the tree grows, the termites gnaw it right up to the branch ends. Thus the tree grows with a hollow trunk. Fortunately the tree is very hardy and can withstand both hollowing out and the nibbling away of all its green shoots. In this respect it resembles the African baobab. The mallee tree is invaded by reptiles, birds and insects, all of which live in the hollow spaces, where the remains of their food and droppings bring the sparse nutritive salts to where there is need of them, in exactly the same way as the larger fauna of the savannah do when they stand in the shade of the trees and, so to speak, leave it all behind them.

The unforgiving conditions leave their mark of the life cycle of the mallee tree. Unlike the African baobab it does not endure for thousands of years. It is simply eroded away, but before that it has produced masses of seeds and sent the genes on a voyage into the future. It has fulfilled its mission and on its journey it has been eaten up inside. The tree, the termites and the animals in the trunk have all developed a complete mutual dependence. It's a package solution, but had it not existed there would have been another. That's the way a true *bricoleur* works.

The wildlife migration – nature's great performance

A gnu calf, a wildebeest, is swept away by the powerful current of the River Mara; it jerks unnaturally, fighting to follow the little flock of pioneers that have thrown themselves over the bluff and down into the muddy eddies of the river. It is picked up again by the current and disappears in an instant. Most of them hold their breath and observe nature's drama in silence from the front row of the stalls. Up comes the head again: is it swimming? Down again, and suddenly the legs are sticking up out of the water, the head turns the wrong way! A crocodile's pointed snout is just discernible with a good grip on the flank. Then it's all over. A split second of eternity. Dramatic and devoid of drama at the same time.

I look around. The flock of spectators is loosely distributed among a good sixty vehicles in various conditions. Several hundred safari tourists from Australia in the south to Iceland in the north have found their way to the annual drama in the Masai Mara. Most of them have paid a small fortune to experience the annual migration and the dramatic scenes when the rivers are crossed. More come every year, and every kind of accommodation is filled to bursting point. The vehicle park gives the lie to the dream of wild, desolate and virgin nature. The disappointment is however, greatest among those who forget that the world is bigger than themselves. Cars or not, the experience isn't really too bad. We are there and we are not there. Life goes on and nature goes her own way.

The zebras are in the front and keep an eye out for the enormous Nile crocodiles, which have taken their places in the restaurant and patiently await service. Small families intermingle and gather into one large family before they move off. Perhaps the pressure from the rear will be too great. It is difficult to see what makes them ignore the obvious danger. Time after time, zebra families slip unscathed through the current. With total consistency the crocodiles go for the small gnu calves. A quick tasty snack. Perhaps the soft young hide offers better access to the fresh meat. An old myth says that crocodiles prefer a rotten, well-hung portion of steak. But even though crocodiles are happy to get hold of carrion, it would be completely illogical if this effective killing machine should prefer meat that had gone off.

In the moments of waiting, there is speculation over why the gnus cannot persuade themselves to cross the dangerous river in one tremendous thrust. The small forays result in high losses, and what are they actually going to do on the opposite bank? They come in small flocks and advance and withdraw; a small flock may follow a family of zebra. The hours pass and the calves go down like dominoes. Every time the same way.

Meanwhile we can follow the ancient lizards coming sneaking through the surface of the water, closer and closer to the wildly splashing, panic-stricken prey. Then the closest sinks inconspicuously down below the surface and suddenly reappears like a fish biting on the hook. If we're really lucky the crocodile's head comes right out of the water and the clicking of a few hundred cameras is sufficient, for the briefest moment, to drown out the grunting of the gnus.

Apart from the close bond between mother and calf, the gnus have no family or herd structure. Those in front lead, and the rest follow in what seems to be the most disorganised and unsystematic behaviour in nature. The same dangerous ford can easily be crossed three or four times by the same animals. Some people believe that the animals can smell rain or certain nutrients in fresh green shoots and therefore navigate by means of the scattered tropical thunderstorms, which certainly seems very unsystematic. It is, however, a fact that the migrating animals follow the overall patterns of precipitation, which say rain in the southern Serengeti in the period from November to March and then in the Masai Mara.

The migrating gnus have come to terms with being nature's enormous lunchbox, and defence and flight instincts are replaced by the mass strategy, "If we are sufficiently numerous, then I'll be the one to survive!" Masses of predators have specialised in gnus, which, on their side, have given up using any further energy on anti-predator behaviour. The individual gnu means nothing. But the species may be said to have had tremendous success.

The crocodiles have no real problem with stuffing themselves in the migration period, which can provide running supplies of food for four five months: this is more than adequate to keep the population plump and well nourished until the game comes past next time. In crocodiles growth is directly related to food intake. When food is scarce the metabolism rate drops completely and growth stops. The crocodile is precisely what it eats.

The guides around us are industriously peddling the myth about the symbiotic relations between the migrating animals. The zebra families move in first, eat the long, dry straw and open the approach to the juicy bottom grasses of the savannah for the gnus. It all fits like a hand in a glove – well, almost.

There is good foundation for the myth, because the gnu ruminates and is dependent on nutrition-rich food. It fills up the relatively small stomach and must then await a lengthy digestive process before it can be filled with any more fuel. The zebra is not a ruminant but it has a long intestine and sends large amounts of food through the system. For this reason the zebra is not as dependent on the quality of the food. Last in line are the small Thompson's gazelles, which crop nutrition-rich herbs. Small antelopes have a large ratio of surface to bodily weight and have a high metabolic rate, which demands nutrition-rich food.

This myth is not wrong, but it isn't quite right either. The romance of nature has taken over, because we have this desperate need to rejoice over nature's formidable powers of adaptation. We see the same myth in numerous animal films and TV programmes. Nature is in a delicate balance, something not to be disturbed. According to the myth every animal has found its niche and the species live in balance with each other, they don't go poaching each other's food. The savannah is never grazed to extinction because each grass-eating species has its respective food niche, which is often described as in contrast to cattle which graze right down to the soil.

But does this also hold good in reality, or are we forgetting our critical rationalism? Why should a zebra prefer dry, nutrition-poor grass? Of course it doesn't! But it can live off it, which is quite a different matter. When the grass is high it is not so clever to stick your head too low down as one never knows what is lying in wait. If one can eat the dry tops from above and, moreover, keep an eye on the surroundings, then one's chances of survival are significantly increased. When the grass is nibbled right down to the soil, the zebras go joyfully amok, tearing into the new grass shoots in fierce competition with the gnus and the other savannah wildlife. The cattle graze right down to the soil just like all the other grass and plant eaters, when conditions permit

The greater the food potential, the greater the success. Wildlife which can adapt, survive – the wildlife of the savannah will die when the savannah disappears. Survival strategies differ from species to species; every animal succeeds with its own. Some are conservative, some are good at adaptation. If they all had the same strategy there would be none left. Exactly the same with humans.

In the late 1660s the Dutch literally threw the Bushmen out of the Cape in

South Africa, after which the latter gave up the newly developed cattle farming, which at that point was a modest 1,500 years old, returned to the Kalahari and resumed desert life. Their culture is therefore far more development-oriented and pragmatic than that of the conservative Masai.

Kierkegaard: when science stares up the arse looking for an intestinal worm

The plant cell is equipped with a stiff cell wall, so that, unlike animal cells, it cannot just putter off round the world, but is good at stacking up. It may find itself standing all its life in the same place, like a dandelion, baobab or grass. On the other hand these species are good at sending off their seeds on a long journey.

Fifty million years ago the plant kingdom expanded with the new, successful family we know as the grass family. Grass produces masses of seeds and numerous variants which are precisely suited to a new soil, other concentrations of salt or water, different conditions of light, in short new ecology. The grasses developed into many species and spread all over the savannah, around the Earth and later to the North and South Poles. Fifty million years later they became the foundation of human civilisation. We learned to cultivate and improve them by breeding, but that is another story.

Parallel with the development of the grasses themselves, came that of the grazers. They developed intestinal systems which were good at wresting from grasses their nutrients. Some acquired long intestines and became effective eating machines, while others developed short, complicated intestinal systems and became ruminants, specialised in quality foods. Some became small agile duiker antelopes with a high metabolic rate, while others became giants like the elephants. In between there are all conceivable sizes. A size for every niche. Parallel with the intestinal system the species developed food-finding strategies. Some became extremely specific, while others became generalists. Some combined grasses with herbs, leaves, bark and one small, grey duiker actually turned its attention to small animals. Behaviour is also part of the picture. The hippopotamus, as already mentioned, has developed an 'easy life' with a low metabolic rate and consumption, while others are constantly on the move in search of food.

Wildlife and plants developed a complicated pattern which has room for all, while at the same time maintaining the dynamic of competition. Part of the pattern is what we prosaically call 'The Great Migration' as described in the previous section. The big connections between the various populations and their environment are normally described as ecological contexts, and ecology is in its pure, original form just that: the description and study of the major contexts in nature.

Gnu, zebra and Thompson's gazelle migrate together in a gradually acknowledged and well-studied pattern between the southern plains of the Serengeti and the Masai Mara. The migration follows the precipitation pattern and the growth of the attractive, nutrition-rich grasses, but many other species, such as the Topi antelope, elephant, Grant's gazelle and birds also migrate. We know their biology and what they eat, but their complex migration patterns have yet to be studied and described, which is possibly connected to the fact that species such as elephant tease the ecologists by using their brains. They remember, think and optimise with the aid of experience. This means they are also highly unpredictable, but more of that anon.

If one imagines that all the grazers are 'settled' or 'resident' then there would only be space enough for relatively few large animals, in fact just as many as can be fed in the dry season. Years of study indicate that the collective gnu population in the Masai Mara–Serengeti would be around 160,000 without migration, while censuses put the migrating population at about 1.5 million in the best years. Wildlife which follow the pattern of the rains, therefore, either optimise the biomass or fill all the niches.

Today the Great Migration of the Masai Mara–Serengeti seems so spectacular that it should be easy enough to notice. Even so we have only really paid it serious attention for a brief fifty years, despite the fact that before that many nature studies and numerous preservation programmes had been undertaken in the same area. European – especially British – nature-protection organisations have been extremely active, particularly when it's a matter of evicting the local human population from valuable nature areas.

One simply had not known what one was supposed to be looking for before a German zoologist, Bernhard Grzimek, and his son Michael flew over the region in order to count the wildlife from their small Cessna. They discovered how the animals migrated, and described the migration patterns, which was later to lead to a new kind of preservation politics Sadly their African adventure ended with Michael's death following a plane crash, but it was immortalised in a world-famous film and book, *Serengeti Shall not Die* that marked a milestone in the natural sciences, in that it was a case of observing everything from a slightly higher perspective. The point is that one sees something quite different from a plane than from a microscope and that each observation can be equally valuable and interesting in its own way. It is only in the twenty-first century that it has occurred to us that nature can be observed to a certain advantage from above. The hitherto dominant view, that nature should be studied in the smallest detail to enlighten us about the totality is called reductionism and has its advantages, but it has proved to be immensely awkward to infer anything from an intestinal worm in the rectum about the migration of the gnu.

Today new ideas and theories about the unity and health of nature appear in

Giraffe, at their most vulnerable.

a steady trickle. It has become acceptable to regard life processes as a cohesive network – an eco-system if you like – even though most theorists take pleasure in continuing to write people out of the system.

One of the best known is the English natural philosopher James Lovelock, who is not content merely with incorporating biomass or the sum of biological life in his mother-Earth theory, best known as the Gaia Theory. According to this there are no processes in nature which fall outside the earthly commonwealth of life, which is in mutual interchange with the physical conditions. By this token the atmosphere can be seen as the skin of the mother organism.

Gaia rebuffs all attacks on her equilibrium. Human beings are a part of this totality – a modest part – but if our activity threatens Gaia, Gaia responds by establishing a new equilibrium, which, naturally, does not adopt an especially human outlook. To put it bluntly, if we don't watch out there will be no place for us. If you cut down to the bone through the slightly coloured filter of mythology and religion there is a great deal of common sense and natural science bound up with Gaia, which is illustrated not least by the present debate on greenhouse gases.

Most of the natural scientific disciplines have never been that good at the overall context ('the big picture') which, to a great extent, has been left to philosophers, humanists – and ecologists – even though one can be superbly in agreement on a suitable, mutual contempt. Good old Søren Kierkegaard was one of the first to start throwing stones, when he wrote, "Scientists immediately begin to

The story of evolution seen from the African savannah

'entertain' with their [endless] 'details'. Now one must travel to Australia, now to the Moon, now down in a Cave under the Earth, now, Damn it, up the Arse – after an Intestinal Worm; now the Telescope must be used, now the Microscope; who in Hell can stand it?!" And the scientist can, according to Kierkegaard, be an outstandingly gifted person, who can explain all nature "but cannot understand himself". So stick that in your pipe and smoke it!

Giraffe science

This morning's beautiful scene with fourteen giraffes drinking was really bad luck. There's no question they were the same fourteen giraffes that have visited us at the camp for the third time in a week. This same week I had explained to our safari guests how modern behavioural research has demonstrated that giraffe groups are very loose associations, which, as a rule only keep together for a few hours at a time. I must have been full of it.

It pays to be cautious with cocksure notions, regardless of what the handbooks come up with. They are often written on each other's backs, repeat the same spelling mistakes, while the more serious mistakes and misunderstandings are summed up as 'facts'. One pleasant exception is Dr Estes's *The Behavior Guide to African Mammals*. Where doubt often takes the form of words, and everything concerned with nature should take doubt into account.

When, later that day, we ran into a new, larger herd of giraffes, I cleverly avoided the giraffes' controversial family life, and kept to the safe story of giraffe kindergartens. Facing the participants and with my back to the animals, here comes the story: the giraffe mother brings a newborn foal into a hiding place where they can familiarise themselves with each other's appearance and scent in peace and quiet: a very widespread strategy among mammals. It often happens that several foals are brought to the same place in the initial weeks when they have difficulty following the adults or fleeing from an attacking predator – hence the concept of giraffe kindergarten!

Merely to turn halfway round in the seat is enough to reveal my reluctant comic mistake. A large giraffe stood leaning over a little foal and petted it head against head. It was difficult to see the foal behind the mother's gigantic head, which in no way obscured the fresh umbilical cord beneath the foal. The foal was newborn, probably some time within the previous twenty-four hours, and thus in no way taken any place whatever.

Now one can hardly reprove giraffes for politically incorrect conduct, but nature can only be described as possibilities and probabilities. Little by little we must accept that behaviour should be regarded as probabilities, but in reality it looks a great deal worse with the exact sciences, when we see things from a giraffe's perspective. Giraffes probably behave the way they should, even though exceptions occur and can be explained, but if new, different possibilities present themselves, not even the probability will be confirmed. So what can one depend on?

In the beautiful mountain reserve Entabeni in South Africa I noted that many of the giraffes have scars on their backs from attacks by lions; one of them was simply missing its entire tail. A local ranger, who also just happened to be my own daughter (her with the audacious questions), recounted that they had, in fact, lost a good many young giraffes to the lions, which had developed a special 'giraffe technique.' Several attackers disturbed the balance of the giraffe by attacking and securing its legs, biting firmly on the tail and simply tipping it over. The unusual element is that scientific probability tells us that the giraffes constitute less than 1 per cent of the lions' diet and that this share consists primarily of foals less than a year old. When giraffes are more than one year old they seldom end up in the bellies of lions. Their superb eyesight detects the danger in time, and should the lion come too close they can deliver a lethal kick. The exceptions do occur, not least, in modern, fenced-in reserves where possibilities of flight are limited.

In Kruger National Park another exception has been created by the expansion in tourism, since the lions have learned how to hunt giraffes out on the asphalt roads, where their enormous, smooth hooves turn them into 'Bambi on the ice'. They become panic-stricken, lose their balance and fall over. Today the giraffe share of the lions' diet has risen to nearly half, thanks to tourism and the excellent roads. Now there exist large areas of the wild without a giraffe population, even though the ecological conditions argue for the contrary (e.g. Selous in Tanzania.) If the lesson from Kruger is learned, one can certainly imagine that the lions, for instance, have developed a technique which has effectively exterminated the giraffe. That would be one of the exceptions where the dynamic between predator and prey has vanished. The imbalance means that one species has been extirpated and another has lost its source of food. This is a far from ideal situation, but parallel with a new dangerous virus which kills its victims. If the virus kills all its victims it will also kill itself.

Nature will probably do perfectly well without giraffes. Other animals will occupy the vacant niche, but the consequences are many and varied. What happens if the giraffes don't gnaw the trees? Will they grow more? Will they lack stimulus? Will there be a change in the dispersal of seeds? Which other animals have advantages and drawbacks? It is not possible to talk about 'negative effects', merely 'effect'. We simply don't know that much about them.

How did a giraffe actually become a giraffe? The Romans, who were familiar with the improvement of domestic animals, believed that, judging by its appearance, this magnificent animal had to be a cross between a camel and a leopard. Therefore they gave it the name *camelopardalis*, which is still part of the giraffe's scientific family name, but what has modern science to say about that?

Doubt is the companion of faith.
Without doubt one falls easily into
the darkness of fanaticism

Concerning giraffes, Lamarck and Darwin

"The giraffes go down into Africa and strive for perfection. An inner striving to be taller and more beautiful and diligent drives development onward, and the acquired characteristics are passed down to the next generation." That was more or less the way my old biology teacher, grinning all over his face, used to present the Frenchman Lamarck to his defenceless students.

One cannot claim that the interpretation was wrong, but if the assertion stands alone it leaves an unjustifiably caricatured image of a formidable man of science, who, on a level with both Darwin and Linnaeus, threw himself into enormous, often overlooked subject areas such as plants, trees and invertebrates. Jean Baptiste Lamarck (1744–1829) is frequently spoken of as Darwin's opposite pole, the man who was wrong and claimed that learned characteristics could be inherited, but in reality he paved the way for Darwin.

Lamarck was digging in the ground around Paris and discovered molluscs in numerous layers. He stated that the molluscs changed gradually down through the layers. A variation in a new layer seemed to continue in a new variant further down etc. This led him to epoch-making conclusions. The presence of the molluscs well inland documented that France had been covered with seawater and therefore the Earth's geological condition was in constant change. The second conclusion maintained that the species were constantly changing through time, as he had observed in the case of the molluscs. In 1801, eight years before Darwin was born, Lamarck wrote that all organic and inorganic conditions were changed through the development of nature and not by the agency of miracles.

Even though the idea of evolution was not common currency in Lamarck's day, he and many contemporary intellectuals had an idea that the giraffe was not created by a miracle, but had slowly developed from something unknown. They knew that the giraffe developed a long neck in the course of time, because it was an advantage. The question was how? What mechanism caused the neck to develop?

In a Danish textbook I read recently that the giraffe has reached its absolute height and that evolution can now concentrate on solving other problems, posed by the fact that it has established its exclusive feeding niche at a height of six metres. This isn't actually the case. In reality, what we know points in the opposite direction. We know that female giraffes prefer the tallest and oldest males. As described earlier, we also know that the largest and strongest males have a decisive advantage in mating. Thus from a theoretical viewpoint evolution pushes the giraffe to new heights, even though this is probably a very gradual process. In practice the small males are rejected, and thus the average height increases.

Theory and practice do not, however, always go hand in hand. If the lions' ability to kill and eat adult giraffes spreads, one might well come to the conclusion that characteristics other than a long neck will become more important. Perhaps it would be an advantage to be a small, more mobile giraffe. Perhaps other colouration would offer better camouflage. In such a case development might proceed in a completely different direction, but how rapidly? We don't know the answer, but we can look more closely at how we human beings have arrived at the idea of an evolution.

Charles Darwin (1809–82) first appeared on the scene when the notion of evolution had already taken a firm grip, and was, by his own admission, inspired by Lamarck and his own grandfather, who had written a bulky literary work on the subject. Darwin harboured a gentleman's genuine respect both for his predecessors and for those contemporary contributors to the theory of evolution, but he still had to 'think himself forward' towards the theory of the development of species. Like his predecessors he lacked the driving mechanism. Darwin's tutor in geology, Charles Lyell had already concluded – like Lamarck – that the force of nature changed the appearance of the world, but now Darwin had to take a seven-league step further, to determine the cause of species development through time.

Grey tree frog.

The man who gave him the idea was the English priest, economist and sociological scholar, Thomas Malthus (1766–1834), whose revolutionary ideas on population development attracted great attention and created a violent political commotion in Europe.

Malthus believed that modern civilisation was facing its greatest challenge hitherto: to provide sufficient food for everyone. He stated that prosperity and public health were rising, while the number of wars and epidemics had declined. The population was increasing dramatically and the greater the increase, the greater the subsequent increase: a vicious cycle. Food production could simply not keep pace, even though it was being made effective and industrialised. The alarming message could be reduced to the following: in an industrial society the population growth would outstrip the growth in food production. Disaster was looming – it was only a question of time.

A little curiously, furthermore, Malthus also concluded that the human being is so sexually fixated that we are not capable of controlling our own procreation, and that regulation will probably lead to sin (adultery etc.). In England Malthus's ideas led to the government attempting to get rid of the 'less fitted' citizens by sending them to the colonies.

Darwin was inspired by the thought of surplus production of people and concluded that all species produce more offspring than is necessary. Accordingly it was obvious that it was predominantly the best-fitted who survived. The small details in a generation are the decisive factor. The worst fall by the wayside and the best have their good characteristics passed down to the next generation. Here was Darwin's missing link: the mechanism which could change the species step by step, the mechanism which could explain evolution.

Both the giraffe and hippopotamus have to give up 50 per cent of their calves to 'nature', lions up to 90 per cent, and many species, termites, water fleas and herrings more than 99 per cent of their offspring. This enormous 'waste' is easy to spot once one looks. The individuals that are best fitted to the present environment are also those which live long enough to breed and leave their imprint on the next generation. The giraffe with the longest neck is perhaps also the best at finding food in the dry season when all the trees, more or less, are stripped of leaves. It transmits its genes onward.

What Darwin observed was, in reality, nature's parallel to his own breeding of pigeons. Darwin bred pigeons and in fact he took this as his starting point when he wrote his epoch-making work, *On the Origin of Species*, which came out in 1859. Like every domestic breeder he picked out the best individuals and mated them with each other until the desired result was achieved. This could be a particularly brilliantly coloured pigeon or an effective carrier bird; in exactly the same way farmers have attempted for thousands of years to breed cows which give a higher milk yield.

Just as the human being artificially selects the best fitted, nature does this in its own natural way. Where the human's selection has a clear purpose, formulated by a human for the human's best interest, nature's selection has no definite purpose, but even so it advances the individual organism's adaptation to the surrounding environment. If the environment changes too fast, the species disappear, others may emerge, and, in the more extreme cases, life itself partly, or even totally, disappears.

Darwin's conclusion was so self-evident that numerous naturalists immediately assimilated it; indeed some were actually said to be kicking themselves at not having 'seen the light' themselves. The naturalist Alfred Wallace arrived at exactly the same idea as Darwin through practical fieldwork in Malaysia, where one of the things he noticed was small variations in some species from island to island, and he concluded that they must have the same origin. Islands, and not least the isolated development of animals and plants on different islands, were, and have continued to be, important arguments and objects of study when it comes to our understanding of the processes of evolution.

Darwin's famous Galapagos finches were gathered from many different islands, without him, however, managing to discover their mutual interconnection. Today they are called Darwin's Finches and are probably the world's most illustrative example of species variation, or, if I might be so bold, species development. Among today's evolution biologists and rabid opponents of the theory of evolution there is, in fact, great discussion of precisely this situation that both Darwin and Wallace described. What is a species? How is one to document that species are split and develop into many new species? How can we document that Darwin's finches come from one and the same species and that they continued to develop into totally new species? Even though one can see that many species are so close to one another, that the connection appears quite self-evident, what is the final, definitive proof? Self-evident is inadequate if this is supposed to be science. Fortunately today we derive considerable assistance from molecular biology, to which I shall return later.

According to Darwin and Wallace – if we are to grant complete respect to the latter gentleman – evolutionary biology can thus maintain that the giraffe with the longest neck was selected because it was slightly easier for it to find food all year round. The best eyesight also offers slightly better protection from predators. Finally, a long tongue can increase the effectiveness of the long neck. These characteristics do not exist, naturally, in isolation, but in the company of numerous other characteristics which are simultaneously weighed in the balance of evolution. Blood must be pumped effectively to the brain. The giraffe must be able to tolerate putting its head down when it is drinking. It must have an effective intake of oxygen, which prevents the overloading of the heart. The pool of characteristics – great and small – is enormous and no one can know in advance which characteristics or which combination of characteristics are the most important. The surrounding environment consisting of climate, water, food and enemies is the decisive factor. Who or which combination is best fitted here and now?

The giraffe's collective characteristics can be seen as a function of its parents' and ancestors' environment. There is no ideal characteristic, there are only the survivors, who transmit their genes onward, with a preponderance of positive characteristics, i.e. characteristics which at that actual moment in time equip the species to the best advantage. An advantage is no guarantee of survival but the statistics are on the side of the advantages which, however, can simply disappear from generation to generation, due to bad luck, for example.

One of nature's most conspicuous characteristics is the brain, which, in certain species – the human being, for example – can ensure the ability to respond to environmental changes. This can be an advantage, but a large brain demands a great deal of energy, both during development and in use. Therefore it is not necessarily the smartest solution. The brain of the giraffe and hippopotamus is not blessed with evolution's great magnanimity, and therefore the behaviour of these species is fairly fixed. They have no strategy to deal with drought, neither have they any means of responding to lions that hunt in unorthodox fashion. We can, however, state that their set of characteristics is reasonably successful, since they are actually still around. The ability to learn or think out strategies is not a precondition of success.

That's merely in our imagination.

Despite resistance from the church, Darwin and the theory of evolution achieved enormous success. There was good common sense, cohesion and logic, and the contemporary enthusiasm for modern natural science and technology provided even more momentum, even though no one was capable of sticking the spade deeper into the ground to uncover the inner mechanisms. Organic processes, genes and molecules had to wait until the next century, but there was no longer any doubt that characteristics were transferred from generation to generation and that nature made its mark on the individual in the same way as in the breeding of domesticated animals.

When a giraffe inherits a long neck we may suppose that it is an advantage caused by something or other. All that it takes is to find said cause, unless it is a reminiscence from a previous stage of development. Nature is still, however, a *bricoleur*, full of strange qualities and appearances, which can be difficult to explain. What is the large insect, the stag beetle, supposed to do with its weaponry? What about the peacock with the giant tail? Why is the hammerhead shark equipped with a completely stupid head with an eye at each end of a rolled-up newspaper? One frequently hears talk of mating advantages, but why so overwhelming? It is however worth mentioning that it is believed that the hammerhead's curious shape gives it 360-degree vision. In every respect a massive advantage, but we cannot draw a definitive conclusion, merely render probable that that's how it all hangs together. And this is true of most of our answers

It took almost 100 years to take the great leap from Darwin to the fundamental genetic mechanisms, but even in Darwin's lifetime the monk Gregor Mendel demonstrated how hereditary characteristics were transferred from generation to generation through sexual reproduction. He experimented with pea plants and showed that the new generation inherited a given characteristic from both parents, but that they had different strengths and could respectively dominate or be repressed. An organism could have visible characteristics, but also be the bearer of other characteristics which were hidden. Darwin, unfortunately, had never heard of Mendel's research, which would certainly have put many pieces in place for the great thinker.

DNA

At the beginning of the twentieth century there was the idea of calling the agents behind the hereditary characteristics genes, and in the course of the

following fifty years we also gradually exposed their biological-chemical structure. To put it briefly, together the genes are called the genome, and this is found hidden in the chromosomes of the cell nuclei. The genome is made of a substance called DNA. Every cell in the organism contains an exact copy of this DNA, which also means that the total of hereditary characteristics can be found in all cells. The DNA 'lump' consists of 50 per cent hereditary materials from each sex, i.e. equal amounts from each of the parents.

Nature, however, has many examples of asexual reproduction. Here the new generation is a copy of the mother organism, a clone, and here the DNA, it goes without saying, does not contain a mixture of the two sexes' hereditary material.

A gene is a stump of DNA which contains the code for a protein. The genes tell the organism which proteins it must produce. Proteins function as hormones and enzymes and many other things. The proteins are responsible for all the bodily functions. They decide whether we are happy or depressed, how we digest food, what colour our eyes are, how good at sport we might be, and whether or not we have a tendency to develop diabetes, cancer, high blood pressure or obesity.

They determine the length of the giraffe's neck, the size of the cervical vertebrae, the ability to take in oxygen and every other characteristic. Now we come to that which changes us from generation to generation. Our genes actually have a slight tendency to change a little, so that a so-called mutation occurs. A mutation brings about a small change in the protein for which the gene in question has the codes. Some mutations seem to be completely neutral, because they do not bring about changes in the protein, while other mutations are so radical that they kill the organism. One of evolution's missed shots.

Some proteins are called regulator proteins, because they regulate the quantities produced of other proteins. They are often capable of shutting down or commencing production of an entire group of proteins simultaneously. If critical mutations occur in the genes which have the actual codes for regulator proteins (regulator genes) the effect can be very extensive or downright catastrophic.

The mutations can affect the composition and thereby the appearance, structure and quantity of the individual protein. When a mutation makes a protein change its composition, it folds itself in a new way which is sufficient to change the function radically. These conditions mean that reality is far more complicated than the classic image of the classic 'genetic model', which says that a gene is equal to a protein, which is equal to a characteristic.

One way in which mutations occur is when our genome (DNA molecule) is copied. In this case an inconceivable number of atoms bound together in the giant molecule are 'read', so that absolutely perfect copies can be made. Mistakes are made. In fact mistakes happen with statistically exact frequency. If one imagines that one has to transfer a series of 100,000 random numbers from one sheet of paper to another in an unbroken sequence, there is a certain probability that even the most painstaking person will make a written error. Something like this happens during the copying process so that mutations occur. There is no other explanation as to why they occur. Is it the *bricoleur* at work, or do the mutations represent the dynamic of nature for both good and evil? We do not know, but they appear with so great a regularity that they are used to temporally determine organisms' sequence of developments. For example, researchers have compared genes for the same characteristics in the human being and the chimpanzee in order to see when we were separated and went our separate ways. This happened, incidentally, around 4.5 million years ago, assuming that we've calibrated the instrument correctly.

There is considerable agreement that mutations are evolution's most essential mechanism, because they represent new possibilities, even though there are masses of, or more accurately, most, missed shots. These mistakes are what we tend to call inherited diseases, but from nature's side, this is merely a part of the surplus production: nature's larder.

Let's leave the giraffe where it is for a moment, and turn our attention to the Canadian brown bear in order to look more closely at how we imagine evolution in action. The omnivorous bear lives in an environment with great rivers, where, through the ages, the salmon have swum upstream in order to spawn. After spawning the salmon dies and it's a no problem for the bears to consume half- or completely dead fish. A mutation may have equipped a single bear with longer claws, or should we call them fish hooks? The bear has thus a real opportunity to prolong the season with fat and nutrient-rich food, because it can now get hold of the fish before they are completely dead. One day this potentially new genetic strain may mate with a bear which has longer arms, faster reaction capability, better eyesight or sharper teeth. Evolution is now in the process of developing a super-fisher, which quickly and easily fills the layers of blubber before the onset of winter: a bear, which, without a doubt, has new advantages in relation to the given environment, compared to the 'old' bears. This assumes, however, that the bear survives long enough to mate and that the new characteristics give its progeny so many advantages that they spread rapidly throughout the population.

The brown bear has acquitted itself well in the northern forest regions and it has gradually spread towards the coasts, the tundra and the polar regions. All species spread towards the boundaries of their existence. On the coast a new type of prey is discovered: the seal, whose fat layers of blubber provide good energy. It is difficult to get close to the seals, and when winter sets in it is quite impossible because nature extends no camouflage to a brown bear. A single mutation creates a white bear – or rather a bear which has lost its colour – a mutation, which in other circumstances would have been lethal, but not here, for suddenly camouflage and environment hang together, and the polar bear's line is under development.

As a matter of fact, the brown bear and the polar bear are the same species. Doubtless they are moving farther and farther away from each other, but the new mutations are not so radical that they cannot mate and give birth to fertile progeny – our definition of a species.

Mutations which change the exterior characteristics are frequent. Other animals such as the mountain hare and arctic fox have followed the same trail. The mountain hare has to protect itself from the arctic fox, but the latter had a new plan underway and developed a white winter coat.

Sceptics at this point will raise the objection: what about the giraffe? Its adaptation is far more complex: how can all these features be developed simultaneously? The answer is really very simple: because evolution is not directional. There is no objective, no decision taken as to whether or not the giraffe should have a long neck. We could just as well have had a zebra-striped, short-necked giraffe with a trunk. The complexity has been under way for a long time, but there are no marked steps; all organisms are complex at all times. Evolution slowly displaces the complexity, and what comes out of it is what suits the actual surrounding environment. An upper lip becomes the stump of a trunk and may improve the sense of smell, but there is no one to say that it concludes by being a complex instrument, the use of which demands an enormous brain. Nature uses what is accessible without any other long-range objective than to optimise the possibilities from generation to generation. Every single generation has had its own mutations, each of which has been assigned a value through evolution. The good, or usable, mutations have spread through the population, and have mated with other good mutations.

One illustrative example is the Australian platypus which is an intermediate form between a mammal and a reptile. It lays eggs like a reptile but the young are suckled by it, fairly certainly directly on the skin of the abdomen since it has never developed nipples. The platypus – and the marsupials – were formerly regarded as underdeveloped mammals, which were cut off from development through continental drift. Neither of these animals is, in any sense, underdeveloped: they are, on the contrary, superbly well-adapted to their present environment. Their ancestors were cut off from the clade (a scientific expression which describes a 'branch' of the 'tree' of life), which turned into the known mammals such as the giraffe, elephant, hippopotamus and the human being, but they have, during exactly the same period of evolution discovered their own path under the influence of the surrounding environment. The environment has thus developed mammals, marsupials and platypuses (monotremes) from the same material.

It is, of course, impossible to say what would have happened if mammals and marsupials had evolved in parallel in the same region. Would one group have 'out-competed' the other? Ultimately evolution has developed bacteria, marsupials, platypuses and people from the same material. Evolution, moreover, has equipped the platypus with a poison gland and a fairly powerful poison. Might this also be a throwback to the reptiles? The same poison is also found in snakes, that are believed to have developed the venom through mutations of the same gene as in the platypus, but completely independent of it. What is interesting is that the same enzyme is to be found in humans; it is simply that we have had no use for developing anything other than ordinary poisonous behaviour. Theoretically, we can rapidly become more venomous.

Intelligent Design or it's a knockout

The story about the platypus prompted a discussion which was only further stoked by the evening's bonfire. After a couple of cold drinks, tongues were loosened and I was somewhat upbraided about the theory of evolution: "… the latest research has clearly documented that many organisms and organs are so complex that it is impossible for them to have developed through evolution." Go for it! After three days with hundreds of examples of the effects of evolution, I have to argue from the beginning all over again. It strikes me that there is nearly always free rein as long as the talk deals with animals and plants, but that legitimate references to the development of human beings clearly

provokes some people. It is difficult to imagine a bacteria cell developing into a human being along the long and complicated pathway of evolution, if one only focuses on the organism, and ignores the forces which drive that development.

The most resolute opponents of the theory of evolution are darned good at putting forward rational arguments in scientific packaging, which affects the doubters because the material is presented pretty inaccessibly. The core argument is nature's complexity. Our organs and the ingenious contexts in which they function are so incredibly complicated that they must have been established or created simultaneously. An eye, for instance, can, however, only function as it was intended to if the cornea, retina, lens and musculature are all present at the same time. One rejects the idea that the individual parts of the eye have developed in several stages, because each of them has a specific function.

The opponents of evolution call their theory 'Intelligent Design' and it is often presented as an utterly new, exciting theory, which gives the theory of evolution a close run for its money. In recent years there has been an attempt to reinforce the theory by accepting certain evolutionary processes, while it is maintained that the crucial contexts were created simultaneously. In this way one escapes being held to account for the situation in which evolutionary features appear quite obvious and unavoidable but retains the right to choose on every shelf from among the examples which can reinforce the theory.

This turn of events means, however, that one has put oneself in a position beyond all the framework of what we normally accept as science. Science is a well-defined concept, which has as a prerequisite an utterly unambiguous theory and practice. A statement is either true or false. Either the theory of evolution is correct as an explanatory principle of the development of species, or it isn't. If there is even a single, well-documented example of intelligent design the whole edifice collapses.

The most noticeable opponent at the evening fire is a fellow scientist who forces me down the path of moderation, for despite all, she is a guest, a friendly creature who would love to unite our points of view. Surely it cannot be so difficult to concede that evolution cannot explain everything, especially not nature's most complicated mechanisms? The arguments are straight from the book. I politely agree that we cannot explain everything, though only because it is an utterly overwhelming task. Think about it, there is a mass of things, which resist explanation, right? Yes and no? In nature there are numerous inexplicable mysteries but when we dig into the essence of things, we discover that evolution functions as an explanatory model. It is far from automatic that the explanation hits the bull's-eye first time round, but we are on the right track and the rational conclusions lead us ever closer to the right answer. Compromise is, unfortunately, not an option.

It's probably just as well for all concerned that darkness cloaks my facial expressions when she reiterates the postulate that the latest research has documented ... the expression is closely related to that classic of television news programmes – 'investigation shows'. The actual statement refers to the book *Darwin's Black Box* written by the American biochemist Michael J. Behe, who is a prominent advocate of Intelligent Design. The book's opening chapter is actually about the eye, but the story is well-garnished with chemical formulae and reactions, which show how the proteins of the eye alter their structure when affected by light particles (photons). It looks tremendously difficult, and it certainly is, but it is merely a description of the mechanisms which research has already mapped out and which are to be found in every textbook for students of life sciences.

Behe puts forward many other arguments to reinforce the idea of a consciously created complexity. He reflects on the difference between simple inorganic chemistry, which goes on around us, and the extraordinarily complicated organic chemistry which goes on in all living organisms. Something must have created this difference. Behe writes and argues well, but this is already familiar material and not new knowledge, let alone science. One is tempted to ask the question: why is it so important for some talented people, who undoubtedly understand all evolution's mechanisms to strangle the theory of evolution?

It is common knowledge that American supporters of Intelligent Design are fighting a tough battle to get their conception onto the school curriculum, so that the prescribed teaching should offer equal emphasis on evolution and creationism. How far they believe in the latter, one can only surmise.

Old wine in new bottles

The discussion continues at breakfast, in the car, at lunch and so it proceeds for most of the week, without my 'facts' bringing us any closer to each other, even though I get an increasing sense of what it fundamentally is that we are discussing. Is it belief, theory or knowledge?

Intelligent Design cannot claim to be a new theory: quite the reverse actually, it is an old belief which could hardly be older. It is, in fact, just as old as the Bible, but the concrete argument was first formulated seven years before the birth of Darwin by the gifted English theologian, priest and philosopher William Paley. The world was also out of joint in Paley's time and he protested against the contemporary heretical ideas about the Earth and the constant changes of life. He put forward exactly the same arguments to be seen today from supporters of Intelligent Design before the theory of evolution was fully developed.

Paley was highly aware of the intellectual discussion taking place in learned and prosperous circles about the probable development of life, which contravened the biblical teaching of God as the creator of all. However, the understanding of the Earth's geological processes was beginning to take root. One could easily see how rivers and glaciers shaped the landscape, how coastlines changed and lakes were formed. The formation of the planet was anything but static, and the leap to the emergence of life became shorter.

It was known how for several thousand years human beings were able to influence and change domesticated animals and plants by selecting animals with particular characteristics for breeding. Lamarck had already observed how organisms had apparently changed through the ages, and, indeed, Darwin's own grandfather had published a work which suggested precisely how living organisms developed. Even though the material was far from generally known at the beginning of the nineteenth century, it was this tidal wave which Paley was attempting to stem before it went too far.

Interest in natural sciences flourished everywhere. The collecting of insects and flowers was almost a gentleman's sport and quite a legitimate occupation for priests with spare time. The new ideas which followed in its wake, however, challenged doctrinaire Christianity and were difficult to handle for provincial clergymen who were, perhaps not interested in those kinds of down-to-earth activities. Paley wanted to anticipate events and gathered all the reasonable arguments and examples in a handbook which village vicars could consult in the event of difficult questions arising in their parish. Paley's works were taken very seriously, were widely disseminated and were obligatory curricular texts at the universities.

The best known of Paley's arguments, that keeps bobbing up like a cork in a rough sea, is the story of the heavenly clock-maker. Paley imagines a man wandering along a path in the countryside. Here he finds a pocket watch and picks it up. After a few minutes' study of the watch's shape, components and function, he is in no doubt that it has actually been made by somebody. It has not come into existence all by itself! The hands, springs and cogwheels are closely interlocked in a unity which must have been created by someone or other. Thus Paley sent a mighty shot across the bows of evolutionary theory, even before it was thought out.

Intelligent Design is not a new idea, which does not necessarily weaken the argumentation, but it is remarkable that here, 200 years later, we are compelled to fight with exactly the same arguments as Darwin: arguments which he chose to reject.

Nobody was offended that I concluded the account of Paley and Darwin by calling Intelligent Design old wine in new bottles, for age does not threaten the arguments, but it was my turn to ask the questions: "Why are a creature's eyes different from species to species? Why have closely related mammals quite different sight. Why are elephants very short-sighted, while giraffes are very long-sighted? Why are sea mammals often equipped with quite different lenses from land mammals?

This discussion taught me that the cocksure are few and far between – the fundamentalists, one might say. Everyone understands the mechanisms of evolutions, and also finds them logical, but even so most people harbour a justifiable doubt about the tiny quark, which, according to atomic physicists, started the entire universe. Excuse me, but where did the quark come from?

It comes as something of a relief that Darwin did not occupy himself with the origin of life, and he saw no contradiction between belief and evolution, even though he ended his life as an atheist. Belief and science are different ways of conceiving of the world, ways which neither exclude nor confirm each other. They are parallel.

Evolutionary leaps at a snail's pace

However, it isn't all that straightforward. The two discussions keep crossing paths: the big one about God and creation, and the somewhat more modest one about evolution. God and Mutation. God or Mutation. I repeat myself. The one excludes the other. Time after time I have to explain that the theory of evolution is in no way concerned with the emergence of life, and has never made a bid to do so. Evolution deals only with the life which has already been created, and its continuing development under the influence of the surrounding environment. Our attempts are exclusively concerned with explaining what we see and apprehend with the other senses. As far as I am concerned, God may very well have created the mechanism which controls the development of the species, as long as we have the mechanism.

My assertion is only reluctantly accepted, for evolution is a notion of the

Hippo, on their way to safer waters.

randomness of nature, and where does that leave God? If everything develops according to chance ... but my rejoinder is that the notion of evolution indicates that the inorganic Earth and all life-forms hang together in a pattern where everything exerts a reciprocal influence on each other. How far this situation has arisen by chance or has been established by higher powers is of no interest to us in this context. The discussion is interesting and in fact I enjoy the arguments which fly around in the safari vehicle as we drive and study our surroundings.

As long as I keep to the unsophisticated adaptation of the giraffes, hippopotami and elephants, and can demonstrate them on the spot, it sort of manages to trundle along, but the further I am compelled down into the evolutionary complexity, the more clearly my problems of explanation appear. How rapidly do the changes take place, and how? Strictly speaking, I don't know too much about that!

Through most of the twentieth century science assumed, like Darwin himself, that it would never be possible to study evolution in nature, because the tempo is far too slow. We have not seen the giraffe become a giraffe. How, then, can we be sure that it happens the way we think it does? Evolution only gives us a good explanation, even though we call it a scientific explanation but we cannot verify the process. We cannot transform a bacterium into a giraffe. Even though we understand the process, we know absolutely nothing of the millions of interim stages which have occurred on the way.

It is assumed that small changes (mutations) in progress in our hereditary material – our genes – carry evolution further, but is that the entire explanation?

Mutations are relatively rare, and their effect is radically different. Therefore there is doubt over this specific situation. How can a long series of mutations in a wide variety of single individuals lead to a complex development in an entire population when, at the same time, we know that many of the mutations are disadvantageous? How do we get from a tiny shrew to a long-necked giraffe through small fortuitous mutations? Biologyspeak takes a very professional turn, when we discuss how small mutations acquire proteins to change form and thus effect. This is heavy stuff, and most drop out, but the point is understandable: that very, very, small modifications may have an enormous significance. Perhaps we should accustom ourselves to the thought that in between the small movements there occur some tremendous leaps.

One would like to know where the mutations come from, and why they appear at all. Good question! If we ignore what we call genetic damage from radiation and the like, we really have no idea. But the most frequent ordinary mutations appear with staggering regularity, as I mentioned in the section about DNA. They occur so regularly in connection with cell division that it resembles an inbuilt mechanism which is so stable that we can date the development of the species and its characteristics by the number of mutations in the specific genes. A quite fortuitous, regular mechanism – doesn't that give one something to speculate about?

We know in addition that extreme environmental effects (radiation, substances injurious to the environment) increase the number of mutations, which can certainly be interpreted as an inbuilt evolutionary mechanism. Mutations are a development potential, and, as such, are neither positive nor negative. That we do not care for mutations which lead to inherited diseases is our problem. Nature is indifferent. Some mutations are beneficial in the current situation and end up in the melting pot of evolution, while others perish because they do not fit into the machinery, or because they are fortuitously devoured before the host has time to pass it on to the next generation.

Furthermore, the lengthy processes of evolution are, notwithstanding, not completely unknown to us, because we have for many years already studied the way bacteria cultures mutate and transform themselves in a laboratory environment. The advantage of observing bacteria is the extremely brief generation span: twenty minutes as opposed to twenty years in the case of the human being. Twenty-four hours in the life of a laboratory bacterium corresponds to 720 years in a human life. We have the possibility therefore of studying how a minute mutation spreads to an entire culture, and, for example, makes them resistant to external attack. We can see how a bacterium learns to resist attack from penicillin through a relatively modest mutation, and thus we understand the evolutionary mechanism which makes bacteria that cause illnesses immune to our intervention. Actually we understand the biochemical processes right down to the smallest atom, and we exploit them in the manufacture of medicine, enzymes and a myriad new products.

Laboratory bacteria are controlled organisms which work for us in the same way as cows and pigs. They are, in principle, bred in the same way as domestic animals, where we have selected the animals or organisms which have precisely the characteristics we desire. The principle is known as artificial selection, as opposed to nature's own selection, natural selection. Even though Darwin used a great deal of the space in *The Origin of Species* on the principles of artificial selection, we are not content with these processes as an exploration of Darwin's most important principle, because they do not illuminate those precise mechanisms in nature which influence evolution.

Now we have arrived at the core of the matter, and I must suffer the indignity of seeing my giraffe story turned against myself. We are, in fact, agreed that the giraffe has an enormous advantage in being alone in its food niche, but I have just demonstrated how many complex solutions there have to be in play before the giraffe can assume its privileged position, haven't I? This is only good-natured winding you up, as we have clarified the question of the many small evolutionary stops and starts in various directions, but popularly said, we would really love to see some examples of how new characteristics spread within the same species in a time span of which we can form an overview. This is precisely the situation which Darwin and his contemporary supporters believed would never be possible.

Darwin's beak

Fortunately it looks as though Darwin was a little too pessimistic, for a century after his death a sturdy married couple went back to the Galapagos Islands to study 'his' finches, and found the answer Darwin had brooded over for the rest of his days. How did evolution function in practice? Which small modifications are required and where do they lead? One thing is to 'make probable' the mechanisms, something I do every day, but documentation is quite another matter.

As previously mentioned Darwin showed no special interest in the birds that, for a hundred years, have been the most famous examples of the development of a new species known as allopatric speciation, and, quite paradoxically, provides today's most crucial proof of the mechanism of evolution.

The small greyish-brown finch is presumed to have strayed from the South American mainland to the Galapagos several million years ago and since spread to all the islands, where it has developed in isolation and consistent with the present biological conditions. As there have been no competing birds it has also been possible for them to evolve several different species on the same island: every species with its own niche, appearance and biology. This has resulted in the fourteen different species that are known collectively as Darwin's finches. The variation in the islands' environments is due to the fact that they are situated in a highly volcanic region, a so-called hot spot. When a new island is formed through volcanic activity, movements in the Earth's crust slowly move the island from the 'hot spot', which produces a new island, and so forth. The islands therefore are of different ages and environments, which have developed in varying expanses of time.

The famous and, not least, hardy, married couple are Rosemary and Peter Grant, and their fantastic story is immortalised in the prize-winning book *The Beak of the Finch* by Jonathan Weiner. For the last four decades or so, the couple has followed the development of the finches from generation to generation in their small tent encampment on the desolate and inaccessible island of Daphne Major in the Galapagos group. Quite simply, they catch the entire population at regular intervals and all individuals are identified; they are also weighed, and the beak is measured in all directions. At the same time the eating habits of the birds are studied and a count made of the number of seeds on the menu. A truly colossal piece of fieldwork.

During an extreme drought in 1977 they asserted that the population sank from 1,500 to 300 individuals, and that the average weight of the surviving population had risen by 5 per cent. There was a corresponding increase in the size of the beak: in length from 10.68 to 11.07mm, and in breadth/depth from 9.42 to 9.96mm. The surviving birds were the largest with the strongest beak. Thus the average of the entire population had increased markedly in a single season. The explanation is called *Tribulus* or goat-head, which is a small creeping plant with yellow flowers, a woody base and beautiful large seeds which are unaffected by drought. The most powerful birds with large beaks were simply better at processing the large, hard seeds, which were the only accessible food.

The drought had another notable effect, in that the relation between the number of males and females was upset from the normal average of 50 per cent of each sex, to a preponderance of five males for every female. The males benefitted by their relatively greater body weight (better metabolic economy?) and it was, to a great extent, the females that had to pay the price of the drought. The unusually violent competition for mating was won by the largest males. The drought forced the population towards larger beaks and bodies. The balance of the whole population was shifted to a measurable extent within a single generation. Completely new knowledge and insight, which, it goes without saying, would have astonished and delighted Darwin.

The Grants' subsequent investigation makes it clear that all extreme weather situations can affect this particular population of finches and that the kind of available food supply is a crucial factor. The climate determines the occurrence

of large seeds, small seeds, soft seeds or many seeds, which determines precisely who manages best. Before the enthusiasm becomes too overwhelming, we ought to ask the question: could it be that in reality the population swings within a more closely defined parameter of tolerance, so that in reality it remains the same, on average, over a longer period?

There has naturally also been criticism of these results which are based on very complicated measuring methods that are difficult to carry out. Is there, for example, certainty of the determination of species? Is it possible the largest birds belong to another species? Can one even determine this in practice? Indeed, one is never allowed to be happy for long at any one time!

Death to the unsuited

Old 'Weiner' is well worn, filled with notes, repairs with tape, underlinings and coffee stains, but I no longer need to lend it out. When discussions require figures we go on the net and look at the most recent updates. Thus it was that in the heat of the battle I discovered that *The American Naturalist* had published a sensational story claiming that there was now, finally, documentation that important evolutionary traits could be developed in under ten years. This was a genuinely welcome news item which proved to deal with the small aquarium fish, the guppy, or millionfish, which can fill every aquarium with fry in only a few months. A research group had put out two populations in the same river in South America where they were isolated from each other by an

insurmountable waterfall. The upper population lived a carefree life without enemies, while the lower population lived an extremely dangerous existence with many enemies. After eight years, or thirty generations (equivalent to around 600 human years), a large number of individuals were collected from each population and a marked difference could be ascertained, in that the fry of the upper population were far fewer in number, but much larger and stronger than those of the lower group. The strategy had changed in the absence of enemies. There was no longer any need for a massive number to ensure survival, and the species can thus aim at producing some larger and more powerful individuals.

The guppy exercise supports and supplements the research of the Grants, and seen together they constitute the best-examined examples of the mechanisms of nature, the small step forward in a definite direction.

Predators, droughts and other extreme climatic conditions seem therefore to be significant environmental factors with a powerful effect on entire populations. Perhaps we may likewise conclude that it is the number of predators, of their collective pressure on the population that makes it necessary for animals such as termites to produce an enormous surplus of offspring which are sent out into the world on a single occasion. Only relatively few get as far as forming a pair and digging themselves in before birds, humans and other mammals have amused themselves with the protein-rich insects. Theoretically they could develop into poisonous, inedible cockroaches the size of rats in a hole in the ground and even further, into the migrating gnus, which every year deliver hundreds of thousands of that year's new calves to predators and the eaters of carrion.

Common to termites and gnus is that their anti-predator strategy is the same as that of the guppy: diversity. If they are enough in number it is easier for the individual to hide inside the multitude and avoid being eaten. This strategy changes under pressure, but how fast will this happen?

Guppy research both in aquariums and natural rivers confirms these same tendencies. Colour, shape and behaviour are affected by the surroundings in the course of a few generations. The effects are so marked that an experienced researcher can identify the predator pressure from the appearance of a specific guppy.

Animal husbandry has provided confirmation that a few generations can alter an animal's appearance radically. In South Africa, for instance, it was thought amusing by some individuals to breed a black impala antelope, just as many zoos displayed white tigers and lions as special attractions. None of these animals would experience favourable conditions in the wild, but, for instance, the dark impalas can be bred if, in each generation, the darkest individuals can be selected and encouraged to mate. As a rule ten generations is enough to 'produce' the desired colour.

Much suggests that natural selection in certain situations of stress can affect a population with a speed approaching that known from the artificial selection of animal husbandry. This does not mean that the results are immediately comparable, as artificial selection only focuses on relatively few, simple characteristics, while the natural selection combines every conceivable factor.

It is easy to imagine how a mutation can provide a bear with longer claws, a finch with a larger beak, a guppy with a greater progeny, and zebra with stripes, but it is far harder to understand why all bears suddenly have long claws, all finches larger beaks, all zebras stripes, etc. A closer look at finches and guppies makes it obvious that one should not content oneself with focusing on selection, but equally on rejection. Nature filters out the worst and the best qualified are permitted to pass on their genes to the next generation. In the case of the ground finches on the Galapagos Islands, 90 per cent of the population died during a drought. And 90 per cent of the survivors were large individuals, which normally only constitute 10 per cent of the population. The whole population, therefore, was pushed quite markedly in the same direction over a mere generation. The less qualified fell by the wayside.

Many species are characterised by an enormous waste per generation, and one must assume that among the survivors in every generation there is a preponderance of usable characteristics, whose share of the overall population rises noticeably with every generation. Small variations which do not make a great difference live on unnoticed through generations until conditions change, after which they either disappear or spread. The most important mechanism of evolution may be best understood as a screening or elimination of the less well qualified.

Appearance is probably an important screening factor. If the zebra didn't have the right stripes, it would get eaten. The marks of the guppy are decisive, both for survival and for mating 'rights', and gazelles cannot dispense with excess heat without a white belly. The human being should at best be either white or black depending on where it finds itself geographically. Organisms which deviate from the norm perish, and the limited tolerance indicates that the genetic processes that change and maintain appearance can proceed relatively quickly, even though we do not know how far this is due to a particularly low complexity with few genes involved, or a massive pressure of the environment. If these processes move too slowly there will, however, be dramatic consequences for the shaping of life, not least for species with long generations. In other words: it's an advantage to be a bacterium.

The theory of evolution is grandiose, flexible, challenging and bloody irritating, for every single time one finds an answer, hundreds (at least!) of new questions arise.

Let us just challenge the above results: finches can develop larger beaks under certain conditions, but the opposite can also happen. Finches with large beaks do less well in years with many small seeds, and evolution forces them in the direction of smaller beaks. Guppies can develop fewer, larger offspring but in other conditions the same fish will swing back to more, smaller young. The latter case can be observed both in aquariums and in the wild. I called that evolution, but is this correct? The individual populations of, respectively, finches and guppies bear apparent genetic qualifications which favour quite different developments. Thus the finch population bears genetic qualifications for both large as small beaks, and, correspondingly, the guppy population for both large and small young. All the possibilities are to be found in the population that thus benefits from its genetic variation. Variation thus protects the whole population from environmental variations. But this is hardly evolution, is it?

All that we know of breeding and natural selection is primarily linked to appearance and well-defined characteristics. But from here it is a colossal leap to the brain's mental processes, and, to understand them requires, in all likelihood, the bringing into play of many genes and mutations.

That's what we think!

Left: A mutually beneficial relationship: oxpeckers remove ticks from a number of African plains animals.

Below: Grounded aerial view.

Left: African monarch butterflies. Spot the male.
Below: A dung beetle with the Waterberg's Hanglip mountain in the background.
Bottom left: A male Lesser-masked weaver struggles with his nest.
Bottom right: People and ants like the sweet nectar from the Sugarbush.

Left: A Ground hornhill snacks on a grasshopper.

Below: The bill of a Scarlet-breasted sunbird is perfectly evolved to fit the narrowest of flowers.

Above: Vultures: the cleaning crew.
Below: Elephants and hippos, downtime at the waterhole.

Above: At over 200 million years old, the crocodile is the oldest hunter in the savannah.

Below: Elephants socializing.

Above: Babies will put their trunk into their mother's mouths to learn about edible foods.

Below: Elephants socializing.

Left: The author with Kofi and parents, three White rhinos.

Below: Sundowners.

The cold universe is more than planets, suns and black holes. We should remember that in certain circumstances elephants and human beings can develop.

I'm here alright, said the elephant.

Cogito ergo sum. I think, therefore I am! If an elephant should utter these words he'd get a long Pinocchio nose alright, for nobody would believe it. Elephants can't really think, can they? But does that mean they don't exist either?

A good 350 years ago the French mathematician and philosopher Descartes pondered what it means to exist, and how existence could be proved. He concluded that the human being exists because he or she is capable of doubting everything and reflecting on his or her own existence. When one thinks it is not as easy to refute one's existence. Therefore to think is the same as to exist. One exists, pure and simple, if one is in a position to reflect on one's existence. From this it follows that, as Descartes the mathematician would express it, other beings do not exist in the same sense as the human being, if we work from the premise that there is no concept of existence in, for example, an elephant's brain. This is something about which we know absolutely nothing.

Using the same, solid argumentation Descartes could also prove the existence of God: God is our notion of the perfect (which we ourselves are not). How can an imperfect human being have a notion of the perfect if

97

it does not exist? Nothing will come of nothing. If we are capable of thinking about the perfect, then it must of course exist. Therefore God must exist.

For Descartes the human being was a rational being, who had the ability to think and thus was distinct from every other living thing. And in that we have believed blindly since the sickly gentleman died of pneumonia in Stockholm.

It is a tad tricky to imagine elephants swaggering round the savannah reflecting on the unbearable lightness of being while they put away a couple of hundred kilos of plant food. Descartes himself was in no doubt: the dumb creatures didn't even have souls, even though he never figured out the purpose of the pineal gland in their brains, for it was precisely the pineal gland which, according to him, was the meeting point of the spiritual, metaphysical world and the material world, also called matter.

This being so, one could certainly torture a pig, or other animals, for the sake of argument without feeling any pangs of conscience. Their cries are due to instinct exclusively. If an animal has no soul, it can, of course, neither reflect on pain, nor know, therefore, that it actually hurts. Today we know better. Nerves and the sensation of pain are there to protect the body. Without a sensation of pain one would not move one's hand away from a glowing hot plate. Soul or not. Today we trust in our senses, perhaps even before anything else.

If we adopt Descartes' dogmas it is precisely because they differentiate us from animals. Where else would we go? It's not too easy asking the animals themselves if they are intelligent, or if they have feelings. But who is really the one with the communication problem? Who can understand whom? Scholars actually try to ask this question quite frequently, but up till now there have been difficulties both with formulating the question and understanding the answer.

Descartes turned his back on empiricism – science based on observations, tests and analysis – and believed, like Plato, that one could 'be content' with letting reason rule. If one possesses that admirable faculty – i.e. reason – one can think one's way to everything without petty-minded differentiation from what our senses might otherwise have told us. This doesn't work any longer. We can quite certainly still discuss Descartes, Plato and many other thinkers, but our development and technological conquests are built to such an extent on a sense-based concept of the world that it cannot be ignored in any context, unless one is a member of the extreme fundamentalist sects. We must first see and then reflect. If we sense, we are.

I have seen elephants larking about in the welcome rain, take mud-baths and sliding around in the mud, scratching themselves on the rump, and I have seen them ramble around disconsolate in prolonged showers, where everything becomes a bit too clammy. I have seen them uproot trees without any obvious purpose, bring luxuriant branches within reach of the smallest, help newborns up on their legs; likewise the adults consistently protect the little ones by positioning themselves between them and alien observers. I have seen them have a chat and take an afternoon nap under an acacia, seen them, radiant with joy, greeting each other with everything that can be found with a trunk and rubbing heads after only a few hours' separation. I have been just as fascinated by the great mammals as dewy-eyed elephant researchers, who inevitably turn into unscientific, doubtful witnesses to the truth after a few months of life with elephants out in the wild. Notice my choice of words. I'm joining the club too!

Maybe we do not go as far as the famous primatologist Dian Fossey, who actually wanted to marry a gorilla, but devotion to these great, magnificent mammals reaches almost human dimensions.

The Kenyan palaeontologist and scientific 'hardliner' Richard Leakey, who has a multitude of books on the origins and development of the human being on his conscience, wrote the book *Wildlife Wars: My Battle to Save Kenya's Elephants* about his time as Director of the Kenya Wildlife Service and his battle against poaching. In the book he describes time spent in the tent camp at Amboseli of the elephant researchers Cynthia Moss and Joyce Poole. In the space of a few days, everything he knew about elephants had a shot fired across its bows. Joyce Poole translated every ear and muscle movement for Leakey with the same accuracy as Jane Goodall had employed in describing the behaviour of chimpanzees. Leakey concludes the chapter with the sentence, "I had become a 'sentimental convert'."

When a human being comes close enough to elephants and great apes, the walls come down and we cease to understand the principle of drawing boundaries. We can see the human features with a human interpretation. Do we actually see what we want to see? Or is the elephant really a thinker?

Leakey's elephant sentimentality has brought me into numerous emotional discussions, but we elephant fans have learned to conceal the gentle side, which seldom compels scientific respect, and conduct the discussions with rational arguments when we meet supporters of so-called 'elephant regulation'.

They say: Elephants destroy trees and bushes and therefore the population should be regulated, so that there is a balance between the size of the population and its toll on the landscape.

We reply: No, elephants keep the savannah open, so it does not become overgrown. In this way the primary production is increased and provides more room for both the elephants and all of the grazers of the savannah. The elephant population regulates itself.

They say: Yes, but the elephants also destroy Africa's thousand-year-old baobab trees in the dry season. These are just as worthy of preservation as the elephants themselves.

We reply: No, trees and elephants are adapted to each other. They have lived together for thousands of years, and even though there were formerly far more elephants, the trees still exist. Show me a tree that has become extinct!

They say: When there are too many elephants, they invade the neighbouring land and destroy the farmers' crops, which have to be compensated for, for example with the income from ivory.

We reply: It costs money to protect the elephants anyway, but with good conservation planning, with buffer zones and hedges, one can avoid most of the problems. The sale of ivory immediately produces poaching, no matter how it is monitored.

This is the way the discussion goes, backwards and forwards, absolutely parallel with the one that is heard year after year at international nature protection conferences without us actually getting any further. A discussion between Cartesians and soppy romantic nature lovers, but it brings us no nearer to the answer as to whether elephants have mental abilities: an answer which proves to be dangerous and opens new doors to unknown territory.

I remember one of the more heated discussions around the bonfire, when I fell over one of our local guides, who was also a regular leader of elephant hunts. And I've got to put it in at this point to be able to remain with the terminology, that I harbour certain prejudices regarding elephant hunters, not least prosperous tourist hunters. I can hardly utter the words without spitting at my PC. Anyway, the poor hunter was definitely not in a majority among my Danish safari guests, so he related how frighteningly animal-like elephants can behave. He had, in fact, often experienced poor elephant calves being abandoned to certain death if they could not keep up with the herd. Definitely not human behaviour! He was correct in his observation, but not in his conclusion, since elephants look after their offspring according to all the rules of that art. The little calf marches along close to its mother, with whom it has bodily contact almost the entire time. If the distance becomes too great, contact is maintained with the trunk, or else the mother pushes its big foot back and feels with its extremely sensitive sole. In addition the entire company can at any time decide everybody's exact position, to the centimetre, with the aid of smell and hearing. Sisters and maternal aunts take a turn and if the little one gets into difficulties there is immediately a flying rescue service, to help it out of a mud pool, move a large branch or hand it down food. Elephants often go on long marches following specific patterns, and the calves have to keep up. If they become ill on the way, they are treated with a great deal of patience. They are supported with the trunk, pushed carefully along, the tempo is reduced and only if it is totally obvious is hope abandoned. At that point, the calf will be so exhausted that it will be unable to stand unaided. The calf is abandoned before it threatens the existence of the rest of the herd. Elephants are also rational.

There are, however, many human parallels to the elephant story. One does not have to go many decades back, before there were among the Inuits the customs of the exposing to the elements, or strangling, of children

who, for one reason or another, had become a burden. If the patriarch – the provider – died in the northerly polar regions, the wife and the mother were forced to make a rational calculation concerning the basis of existence of an entire family: which children could fend for themselves and what hunt surplus they could produce. Part of the story is the fact that the most skilful of the boys could go out hunting alone from

the age of six to eight, and many were competent, experienced hunters at ten to twelve years of age. The children for whom there was no food had to die. In most cases the mother simply smothered the children. I do not think there is a great difference in maternal feelings between then and now.

While I'm writing this the Danish Sunday newspaper *Søndagsavisen* furnishes a picture of a two-year-old Chinese child who, despite the winter cold, was shackled with a chain to a lamppost, every day, while the father was at work. Two years old, by the way!

A nose for elephants

In the spring of 2009, after a break of some years, we started again to experiment with 'walking safari' in the Masai Mara. In principle this involves walking the savannah and looking at what turns up. As a rule we see many antelopes, birds and giraffes, but it does happen that we get to see elephants, buffalo and maybe lions on the horizon. This is, however, far from being as dramatic as it sounds, but safety precautions must to be taken, really in the same way as if one was bathing in the North Sea. Some ground rules have to be observed if one really wants to get back to the beach safe and sound. On a walking safari the most important rule is to keep to the open spaces where one has a clear view of everything around one. It is always the local Masai who are responsible for safety while I look after the talking. It is difficult to be responsible for both at the same time, so the division of labour is appropriate.

Suddenly the small Masai, who has assumed the 'guide name' Peter, stops and explains that there are elephants in the bushes. He can smell fresh, broken-off acacia stems and concludes therefore that there must be elephants. He is, of course, absolutely correct. I try to catch the scent, but I don't succeed in registering the actual scent of wood, even though I am otherwise quite proud of my abilities in that respect. My lack of success can hardly be attributed to the size of the organ in question.

Even though we are capable of training our noses, we miss a colossal universe of smells as a result of evolution's many compromises. Our early forefathers needed to climb trees, which made necessary the development of stereo vision. The snout was in the way and had to go. The shape of the face became flat, and thus the nose shrank and with that we lost a mass of olfactory receptors. Later came walking and talking. The nasal cavity was halved, which did not exactly increase the ability to register smells. We do, however, have the option of training most of the senses to a much higher level than is the average.

The elephants in the background transmit several powerful warning signals, which is somewhat curious, considering they are over a kilometre away. Peter reckons that something else must have spooked them, but I think it's us. The wind actually bears in our direction, so the elephants cannot pick up the scent.

They can hear us and one sense is hardly enough to establish our identity as harmless tourists. Everything unknown is by definition threatening, until proved to the contrary. The same attitude was surely dominant with our ancestors.

There is no reason to be nervous of elephants when in the company of Masai: they have learned, like the lions, that small, scurrying men in red are to be avoided. It is very seldom that Masai attack elephants, though there are examples of young warriors gratuitously throwing spears at elephants, and there must undoubtedly have been numerous earlier episodes where the two groups have run afoul of each other, and killing followed killing. If an elephant kills a Masai it has signed its own death warrant. And they know it.

The elephant's eyesight is poor, but its sense of smell is unsurpassed. Its enormous trunk, nasal cavity and pharynx constitute a real smell computer. They can detect the smell of a single molecule. The number and speed of the smell molecules provide the elephant with a quite complete image of the encroacher: species, size, number and speed. In a sense elephants see with their trunk. An elephant's sense of smell thus fulfils the same purpose as our sense of sight. All senses end with small signals – or small impacts – on a nerve end which goes directly to the brain, where the necessary images are formed. All animals, including the human being, are in possession of various combinations of senses, which finally end in a highly usable image in the brain, irrespective of which sense seems to be predominant.

The idea of the eye as a particularly prominent and valuable sense is a human error of inference. It can be traced all the way back to the ancient Greeks who believed that sight was a person's foremost sense. Both Plato and Aristotle write of the eye as the "mirror of the soul" and in the first lines of his *Metaphysics* the latter links the human desire for wisdom together with the most important source of it: sight. For Aristotle wisdom came through the eye which could see and read, but today, 2,500 years later, we are slowly beginning to understand that the senses are to be found in the brain, and not in the more or less important receptors which relay the signals. We are simply so locked into our concept of the eye's formation of images that we find it hard to understand that usable images can be created by other senses. We cannot, of course, know how these images appear, but they do work. As a matter of curiosity it might be mentioned that researchers work on stimulating blind people's other senses to create a form of visual images to compensate for those of sight.

After all, the senses are there to ensure the continued journey of our genes, which in a more down-to-earth sense means that the senses see to it that food is put on the table and that one does not end up in the stew-pot.

Along the way, Peter and I discuss the uneasy elephants. In theory there may well be lions in the vicinity that have unsettled them, which would spice up the walk to a considerable extent. However, I keep my own counsel. Peter remembers that elephants often graze quite peacefully, side by side with cattle, while on other occasions they can behave with remarkable aggression. Together we can draw the obvious conclusion: elephants are frightened of the unknown, that which they can only ascertain through hearing, but cannot see, or rather smell. The sense of smell supplements their hearing, and together these senses create the necessary image which determines the elephant's behaviour.

We have a definite feeling that the elephants in the Masai Mara recognise a safari vehicle, and probably also the individual cars, together with the drivers and guides who regularly drive them. Therefore one can experience them passing less than five metres from the cars and even poking their trunks through windows or other openings. The fact that we are now approaching on foot is something that will take some time getting used to. In the meantime we should make sure that we have the wind behind us.

The trunk line through evolution

The experienced safari guide has a number of classic 'back-ups' for when the day's experiences don't live up to expectations. He drives to the nearest rock and waits. Shortly afterwards a small animal arrives which looks most like an over-sized hamster or guinea pig. This is a rock hyrax, or *dassie*. A moment later we have the company of the rest of the family, which peep out from holes and fissures. The full-grown rock hyrax is the size of a hare and its nearest living relative is actually the elephant. The connection between the small rodent-like animal and the elephant is not immediately apparent, and such information often seems to blanket the company with a silent scepticism. Now he's really laying it on with a trowel!

The difference in size has, however, no significance to the relation between the two animals but makes one think of their common ancestor, which is a kind of shrew. Many different lines of development have emanated from this ancestor, some of which have become great elephants, while others have developed into small rock hyraxes. A third line of development has returned to the sea and has become the sea cow (also known as manatee or dugong). Large and small species have enjoyed success each in their particular place, their niche in nature.

When I tell this story people often seem to detect a rudimentary trunk in the rock hyrax's snout, which, however does not turn out to be the case on closer inspection. However, what one cannot see is that the rock hyrax actually has the rudiments of tusks in the upper jaw. The story of elephant evolution is the story of the evolution of the trunk. The first elephants entered the world stage about 35–40 million years ago, as far as large animals with tusks and a small rudimentary stump of a trunk can be called elephants. We begin to talk about elephants with the prehistoric *Palaeomastodon*, but the discussion about when an elephant is an elephant, is, in principle, the same as in the case of when a human being is a human being. When are our ancestors apes? When are they human beings? And what qualifies these categories? Is the earliest upright ape-like being, called *Australopithecus*, a human being? Is the Neanderthal? How much trunk does it take to make an elephant?

The elephant's shrew-like ancestor may have descended from marsupials, which may descend from reptiles, which descend from ...

The elephant has developed successfully throughout history. It grew large, in fierce competition with the grazers of the savannah, but gradually forced its rivals out of the running through the development of an extraordinarily long digestive tract which permitted it to subsist on plant material of the poorest quality. The trunk kept pace with this development: it became longer and more powerful and developed gradually into the utterly unique universal instrument we know today.

Like human beings, elephants adapted to life in most of the Earth's regions and climatic zones. Like other animals they explored the boundary of their existence, right up until the ice front where the great glacier landscapes stopped them. Here they developed the thick pelts needed to occupy the last accessible regions. The interim result was a range of species, each with a markedly distinct appearance. The various species occupied quite different niches, and when circumstances changed, so did the elephant – or it died out. The story of the elephant contains a series of ups and downs and thanks to the discovery of ancient bones, skin and entire animals in bogs, tundra and caves we know quite a few of them.

During the Stone Age at the close of the last ice age, a good 10,000 years ago, there were still five species extant: the mammoth, the stegodont, the mastodon and the Indian and African elephant. The mammoth, Indian and African elephant (have) lived simultaneously on the African continent whence the first two later disappeared. We do not know the cause of this, but we can state that the Indian elephant continues to thrive throughout most of the Far East.

The African elephant originally lived in the forests of Africa, but when the other species disappeared from the savannah the niche became vacant, and the species spread from the forests out into the open regions where it gradually adapted to the new environment.

Much suggests that it is not only backwards that one should go for the number of elephant species. In recent years elephant researchers have actually demonstrated that the African elephant is in reality two species: the original forest elephant and the 'new' savannah elephant. This actually fits rather well with the various theories of species development, in which a species spreads as far as the extreme limits of its natural areas of distribution and here undergoes new adaptations. If the forest elephant lives satisfactorily in its niche; there is no need to abandon it, but, on the other hand, such success will compel individuals to range farther and farther afield to the periphery.

Even though the savannah elephant appears to be considerably larger than the forest elephant, researchers must have recourse to DNA analysis in order to establish the difference, and something suggests that there still occurs a certain hybridisation, even though each animal appears to confine itself to its respective habitat.

The two species do not constitute two physically separate populations, but constantly cross each other's habitat. Therefore the splitting into two species only occurs slowly, and it might be foreseeable that the forest and savannah elephants tend to mix characteristics to such an extent that they end up as a distinct third species. It is precisely here that various elephant researchers, systematists, evolutionary biologists and theoreticians clash in innumerable discussions of how species do or don't arise, and what, exactly, constitutes a species. DNA tests on the elephants show two species, but if they can mate with each other and have fertile offspring, then they are not really two species, are they?

Discussion of species flourishes everywhere, where two or more species or sub-species seem to be close. How does one demonstrate that they have the same origin? The latest genetic studies of African giraffes indicate six separate species, which seem to diverge increasingly from each other in a genetic sense. The genetic picture seems to confirm our suppositions about the splitting and formation of species, but this transition is apparently a smooth, lengthy process which continues to withhold any clear answer to the question: when is a new species formed and from what?

Evolution works because the individual species challenge their own geographical frontiers both horizontally and vertically and because the surrounding climate oscillates violently. Many mechanisms function simultaneously and contribute to the differentiation of the individual species at different rates.

The human being and the elephant have developed side by side for the last 50 million years, but during the preceding hundreds of millions of years we constituted a common line. The fellowship came to an end in the first successful period of the mammals, and when the first elephant-like animals appeared on the Earth's grass plains, we were still crawling around in trees like small semi-apes with flat faces and long tails. For 20–30 million years, the apes developed many new lines and species, exactly like the elephant. Both groups ranged farther and farther from their original habitat in Central Africa and changed along the way, keeping pace with the new climatic challenges which had to be met successfully.

Around 14 million years ago there emerged the line of development among the apes which today we call the great apes or the man apes (sometimes known as hominids). Some wandered out of Africa and developed into the red-haired orang-utan which lives today in Indonesia and Malaysia. The others, and this includes the ancestor of human beings, continued together in Central Africa until the point of development when the gorilla separated off. Henceforth we switched tracks together with the genes which were to become our own era's two types of chimpanzee. We accompanied the chimpanzee until about 4.5 million years ago, and thus we are the chimpanzee's closest relative – closer than the gorilla's – if you want to see it that way!

From the moment when we left the elephant's line of development, we shared a history of development and genes with the chimpanzee for nearly 50 million years until we started our own new line and repeated the whole developmental history of the elephant, the gorilla, the giraffe and bacteria, in which a new species occupies new niches and develops itself in various directions. Just like the elephant, we wandered around the planet and developed new species. The new species were due either to the fact that prehistoric man moved towards colder, hotter, drier, wetter or higher regions, or that the Earth's general climate changed.

We can follow both the gradual development of our ancestors and their migration around the planet, even though the picture is far from complete. They wandered along rivers, lakes and coastlines. The journey was determined by fresh water and food, but they all ended up in evolution's blind alley, just like the lizards and many of the ancestors of the elephant.

All the other branchings-off of our line of development are extinct. We know that we have lived parallel with the Neanderthals and possibly even with some small human beings found on the island of Flores in Indonesia, while we – *Homo sapiens sapiens* – are, in consequence, the pitiful remains of all the lines of human development and thus evidence that new species and new lines of development emerge while others disappear. Life adapts to the planet's changing conditions, but the less comfortable thought is that the existence of the human being is apparently not a precondition for life.

Speech – the brain's gift

When the water holes of the Masai Mara dry up, the elephants drink from the river. Hardly a day goes by when we do not have a visit from a family group at the bush camp. They stand on the opposite bank, suck the water up their trunk and spray it alternately in the mouth and over their backs. The drops lie like dark stains on the grey, folded skin. They seldom go out into the water, even though, once in a while this happens after sundown. The river is the territory of the crocodiles and hippopotami, and should only be disturbed if absolutely necessary. After watering, it's time for the daily dust-bath. Dry earth is gathered with the trunk and gets shot up like the water before it over the back where it coalesces as an impervious brown layer which provides a certain protection against bugs and ticks, even though this solution to the problem cannot possibly compete against a good old-fashioned mud-bath, where thick cakes of the stuff remain attached to the skin.

It can happen that several herds of elephants come simultaneously, as if by appointment, but the various kinds of meeting differ. The meeting between family groups which consist of related females and their offspring is always very affectionate. There must be greetings, forehead to forehead, cheek to cheek and the ears flap, while the trunks are entwined. The closer the two groups are related, the greater the devotedness. All the elephants within a wide radius of many hundreds of kilometres are mutually acquainted and highly conscious of the kinship structure. If a single male or a small bachelor group turns up at the same time as a family group, they remain on the periphery of the family meetings without making any closer contact.

From birth to death the elephant's trunk is active round the clock, but only the fewest 'have a nose' for this unique universal appliance, which evolution has developed from a floppy stump of a fusion of labial and nasal skin to one of nature's most sophisticated organs, which far exceeds the eye in complexity.

The trunk is unique in so many respects that one is tempted to fall straight into the trap of thinking of it as an evolutionary end-product, which, naturally, is a load of nonsense. Even though the earliest trunks were primitive in comparison with the present variety, each one represents a usable adaptation to the contemporary conditions in harmony with the other characteristics. The combined characteristics of the modern trunk were, in earlier elephants, merely distributed in a different way, but one cannot, of course, rule out the possibility that it is the trunk of the present-day elephants which has made a crucial difference. The others are no longer with us!

We do not know a great deal about the function of the earliest trunks but our physiological knowledge gives a good indication. The early trunks gave the prehistoric elephant a large nostril, which must be supposed to have been filled with olfactory receptors. The sense of smell has thus had a high priority even from the earliest stages of the elephant's development. The large nostril presumably also played a critical role in connection with temperature regulation, as one can see in numerous present-day species, which can lower the body temperature through the regulation of air currents past numerous surface blood vessels in the nostril. It was one of these functions that our ancestors in the trees lost with the flattening of the face, which made possible stereo vision and the conception of space. With the elephant the cooling function has been partly taken over or supplemented by the large ears, which are provided on the back with a network of surface blood vessels. They do therefore have several reasons to flap their ears.

When I sit in an open safari vehicle in front of a group of elephants and the incessant clicking of cameras finally dies away I often come up with a brief suggestion: follow the movements of a trunk for five minutes and learn a deal about the elephant. You will experience the trunk wrapping itself around a branch or a bark stump, which is then torn off and eaten; the tip of the trunk curls around a shrub which is pushed with the foot after which the earth is knocked off against the knee; the two outer fingers of the trunk which are more sensitive than our fingers, pluck an attractive flower; all while the trunk is in endless contact with the other individuals, the calves in particular. Every so often the trunk reaches up into the air in order to monitor the incoming smells.

The trunk contains around 60,000 muscles, and even though there is some disagreement over the precise number there are many hundred more than in the human body. A single blow of the trunk can kill a lion on the spot. The trunk, to put it briefly, is a complicated bundle of muscles without a single bone. Thus the elephant has overcome the traditional limits on movement of the tendons and joints with which all the other moveable parts are handicapped. There is, so to speak, no turn nor movement which an elephant cannot execute, from the powerful, violent blows to the finest and most sensitive contact. All the muscles are, naturally, directly controlled by the brain, which has the impressive weight of five kilos. The size of the brain is hardly a measurement which serves as an evaluation of 'mental abilities' but there is no doubt that the trunk and the brain have developed in parallel, because the complex movement structure demands a considerable brain capacity. It is also evident that it is precisely the movement centre which is the most markedly developed part of the elephant's brain, together with the centres of hearing and smell. Theoretically such a large brain is not necessary to control the elephant's body, and one may therefore speculate what other characteristics the elephant's brain has equipped the animal with. They have survived the selection and rejection processes of evolution in the same way as the modern human being. Is this because they are cleverer, more intelligent, better able to adapt?

Once I was witness to what cinematographers would call a 'crowd scene' in the Masai Mara. We had followed a mighty herd of elephants, numbering about a hundred individuals for a good hour. They proceeded calmly, grazing, but seemed very focused and directional. A hundred elephants is a rare sight anywhere in Africa, and no elephant herd of that size stays together for very long. They would simply destroy their own basis of existence.

I can, therefore, state with certainty that this was an assembly of five to ten elephant families, probably all interrelated and participating in an annual or otherwise regular meeting.

We drove on and encountered a corresponding herd. This was so unusual that some of the tourists in the car opined that it 'must absolutely' be the same herd, heedless of the fact that we had pulled away from the first herd in a straight line. It was another herd. They were also off to a meeting or pow-wow – a word we have borrowed from the Native American gathering of the tribes.

We had a clear sense that the new herd was moving in the direction of the other herd about thirty kilometres away. After we had photographed a pair of lions, we went back to following the elephants. From a ridge we could clearly observe the two herds heading directly for each other without eye contact.

It was clear that the groups were in some kind of contact with each other, even though they could not see each other. Elephants communicate by means of low-frequency sounds, a kind of stomach rumble or throat noise which is produced without vocal cords and which are so deep that they are inaudible to the human ear. Now there is a limit to how far elephant's hearing can reach, but the sound is transmitted much farther through the ground, and elephants can actually detect seismic waves with the sensitive pads of their feet, and thus communicate over extremely long distances. It has been demonstrated that the two fingers of the trunk also contain highly sensitive cells, which can register vibrations in the ground. The elephant simply lays its trunk on the ground – pushes the plug in – and picks up the oscillations in the low region (infrasound) between 15 and 35 Hz. Fortunately we cannot hear these oscillations, for there is nothing wrong with the volume. It measures between 70 and 100 dB, which corresponds roughly to the sound of a jet engine.

It may well be that one should take care with the big words, but the elephant summit was really fantastic, because the experience was not merely an enormous herd of elephants nor a rarity because one sits back with a pleasant feeling of having been given insight into something of consequence, the scale of which one does not completely understand.

The clan meeting takes place at regular intervals, even if they become fewer and fewer, and smaller and smaller. The family groups consist of a leading female, a matriarch, her sisters and cousins and their children. When a group has reached a certain size, it splits into two, which, as a rule, spend up to half of their time together. The dividing up occurs when the group has reached ten to twelve individuals, depending on the amount of food. In this way the elephants constantly adapt to the ecological situation, if this is possible. In very lush areas of Africa the herds will typically be larger than in arid regions.

The groups which have just split are closely linked and the ties are not loosened until these groups split next time, and thus the elephant families develop into large clans which have all known each other their entire lives. In Botswana it is still possible to experience clan meetings of up to 500 elephants, and formerly there were reports of gatherings of up to a thousand. Even the related males turn up to these pow-wows, the precise significance of which actually remains unknown.

The elephant's communicative skills mean that related family groups can search for food a long way from each other while at the same time maintaining relatively close contact. They are probably no farther away than they can communicate, or, also, they have simply arranged the next meeting place,

A leopard hangs out.

something for which there are many indications. Even after the briefest of separations, they meet with a massive outpouring of affection, as we observed at the bush camp. Trunks are mutually intertwined, there is nudging, pushing and fondling so that it is pure pleasure. Elephant researcher Cynthia Moss writes that, even in her most pig-headedly scientific moments she is not for an instant in doubt that elephants display real joy. Once one has seen it, it is extremely difficult not to concede she is right.

One of the places in Africa where it is possible to study the various meetings of elephants is at the water holes of Etosha National Park (Namibia), where a few hours of waiting will be rewarded with various elephant meetings. As the water holes are relatively few in this bone-dry environment, many families will meet at the same holes. The family relationships are clearly apparent. All close family members get a hug, whereas distantly related groups must content themselves with drinking separately, even though one can see individual group members, that can obviously recall a certain kinship, meet and greet each other. The elephant's ability to communicate via infrasound is far from unique in nature. Killer whales, the smaller dolphins and other whales communicate with relatively low-frequency sounds, which travel five to ten times faster through water than though air. Sound waves are picked up by the jaw bone and the fatty marrow at the very back of the jaw which are set vibrating and transmit the sound onward to the middle ear, inner ear and audial nerve. In mammals the bones of the ear are actually stumps of bone from the jaw, which are patched together into the malleus, incus and stapes.

Sound can, in principle, reach the individual in numerous ways. Deaf people, for instance, lay their hands on a hi-fi system and thus have an experience of music. The main reason is, that one signal or another is transmitted to the brain and is there decoded.

We can fool elephants, apes and dolphins with recordings of their own sounds and thus study how they react to their own signals, which has led, for example to a 'translation' of a large number of ape signals, but there is still an infinitude of things to discover. Were one to subject a Japanese, Chinese or Arab to a marathon performance of *The Ring of the Nibelung*, they would surely suffer a nervous breakdown in the course of a few hours. By the same token there are others who pay money just to experience it: Vivaldi's music is only beautiful if one lives in a culture where it is conceived of as such, though here there are exceptions. In Copenhagen people have begun to play classical music in publically accessible areas where one does not want drug addicts – who apparently can't endure it.

The point is that communication in nature is far more complex than was previously imagined, and it certainly cannot be measured against only a human scale. We are inclined to look for vocal cords, special speech genes or the hyoid bones of extinct species, when we wish to evaluate the communication skills of a species. The ability to communicate, however, hardly resides in the vocal cords, but in the brain. What is crucial has to be to what extent the brain is able to communicate, and how much benefit it receives from communication.

In the previously mentioned book *Wildlife Wars: My Battle to Save Kenya's Elephants*, Leakey describes an episode in Kruger National Park in South Africa where the elephants reacted to a helicopter with apparently unmotivated, hysterical terror. As soon as groups of elephants pick up the sound of a helicopter they flee trumpeting headlong. Panic is a fairly atypical behaviour, but in this case it is attributed to the fact that the 'regulation', or culling, of the elephant population was previously carried out with the aid of helicopters. The strategy used worked on rounding up entire elephant families and shooting down all the individuals so that survivors would be unable to pass on these unpleasant experiences to the rest of the population. The strategy was not successful: the murdered elephants apparently managed to send the message on and the experience has now been transmitted to thousands of elephants in the region.

It is assumed that the elephants can send trampling signals over long distances as a supplement to the low-frequency 'growling'.

In the course of research into the elephant's sonic communication it has become apparent that males and females have different languages, or, at least that the language of the females is far more complex than that of the males. There is a certain logic behind this observation. The young males actually abandon the family at the onset of puberty when thirteen or fourteen years old and live the rest of their lives partly alone and partly together with other males, where they principally fend for themselves and have no great need to communicate.

The females, on the other hand, remain with the family group, where there are a great number of tasks which demand intensive communication. The care for and upbringing of the young is shared out among the females. 'Parental care' is a demanding task and the entire group is highly dependent on the alpha female and the experience of the eldest. Careful consideration must be given to the length of the day's march, choice of food, water and current meteorological conditions. The capable females will adjust the progress of the whole group according to the weather conditions. They know, for example, that the sound of thunder means rain and thus green grass. In other words elephants are able to predict a development, a characteristic which is otherwise considered exclusive to mankind. The females learn throughout their whole lives, and the eldest communicate all their experience to the youngest, who will later take over the leadership of a family.

Sound is, however, far from the elephant's only means of communication. Both the sense of smell and that of touch are well developed and play an enormous communicative role in most animals. Numerous behavioural studies have even revealed a very deliberate use of these possibilities thanks to the large brain. Elephant researchers have described and interpreted up to three hundred visual signs, attitudes and movements which appear to be standard in the elephant kingdom.

We still do not know a great deal about the elephant's thought processes, but it appears that it communicates with all the means at its disposal. The large brain has simply utilised all stomach and trunk sounds, trampling, smells and attitude to dispatch the necessary messages.

Even though we do not know so much about what it uses all this communication for, there is no doubt that the elephant brain is used diligently all through life. It accumulates tremendous knowledge about the savannah's food sources, dangers and weather conditions, and it apparently remembers everything. This is true not least of dangerous situations. It is not necessary to read too many old accounts from Africa to establish that any elephant hunter's nightmare scenario is meeting an elephant he has previously harmed or hunted.

Ecological elephants

When the grass grows green and lush, then there is tranquillity among the elephants. We often drive to a position a hundred metres in front of them and await their relaxed passing two metres from the car. They graze peacefully yet remain constantly in motion, even though, strictly speaking, this is unnecessary, for there is an abundance of food. They move past us clustered densely together, sniffing yet seldom touching the safari vehicle. We have plenty of time to observe the pachyderms, and can, as a rule, see every single wrinkle in the grey skin. The newborn calves have to keep in the background, on the other side of the adults, but once in a while they drop back a couple of paces and run for safety, while the little trunk whirls around in uncontrolled movements.

When a clump of grass or a flower is torn off with the trunk, the elephant simultaneously takes a couple of steps before the story repeats itself. Often the elephant grips the roots with its feet, so it only removes the top. Apart from the impression of the feet there is almost no trace of the group in the grass. No devastation, no overgrazing. The constant movement means that every spot the trunk touches is only plucked once. Seen with Darwin's eyes the elephant's behaviour is utterly rational. As long as the grazing does not go right down to the ground, production continues, to the joy of the elephant and all other wildlife, whose grazing also stimulates growth, just as mowing a couple of times a week will stimulate a lawn.

When the dry season sets in, the elephants range over long distances in forced marches. If we want to observe them at leisure we have to find them in the thicket, where the trail of severely damaged bushes and trees is easy to follow. The light hues from the striped stumps stand out boldly from nature's other colours. Broken branches, fallen trees and flayed pieces of bark hang everywhere like garlands. The destruction seems massive in relation to the small quantities of food that the elephants actually consume on their way. At first this behaviour strikes one as irrational compared to the cautious, ecological interaction with nature in the rainy season.

The damage is, however, not as massive as it first appears. Many of the stems of the small bushes prove, on closer inspection to be disproportionately thick. They are trees which have survived numerous elephant visits and developed into bonsais fashioned by nature itself. Trees and bushes have developed masses of survival strategies and can often live through even the most massive attacks. The dynamic between plants and plant-eaters functions along the same lines as that between predators and their prey. Both species become better and better at managing in the actual context. It is not as easy as one might think to get rid of the growths and the ravages of the elephants can certainly not stand comparison with mankind's destructive felling for firewood.

The elephant, however, does not only ravage bushes and trees in the dry season. In periods of great, lush abundance one may encounter elephants setting their foreheads against one of the larger trees and knock it to the ground without bothering to eat it. This is precisely the situation which has given the elephant an undeserved reputation as a savage engine of destruction.

Every time an elephant succeeds in destroying a tree or just sets back its growth a little, it creates a new spot, a small open field, where its primary food, grass, can thrive. Grass, in its various forms grows much faster when it is not in the shadow of trees, where it also has to compete for the precious water. The savannah's accessible source of food – i.e. that which can be digested immediately by most of the grazers – is increased many times when a tree is toppled. The elephants clear the field, just as peasants have done for ten thousand years. Only elephants have been doing so for a couple of million years.

Trees are, and will remain, a secondary source of food and the elephant's secret weapon in the fight for survival against the savannah's grazing antelopes. However, this is not the whole story of the elephants' interaction with trees, for the savannah elephants have spread beyond the periphery of their natural habitat, and henceforth they must adapt to new ecological conditions.

In certain places in Africa the savannah elephant has actually begun to put marginal land to use. It has simply struck camp and headed for the arid, burning deserts of West Africa and Namibia. Here the family herds are small, and their home range is enormous. One family's home territory may well be 11–12,000 square kilometres, within which they know every watering hole and its annual yield. The small herds have to move constantly between the desert oases. They can be found in dried-up riverbeds where they dig for water and eat solitary trees with long roots, and around the permanent watering holes which are not always as permanent as one might wish. So, always on the move after food and water. Elephants here have dropped the destructive behaviour and never destroy a tree or a bush, unless there is life-threatening drought. Food is very limited, and it is a case of keeping everything alive. They nip at the trees and then move on without damaging the individual growths as if they are thoroughly aware and conscious of the tolerance of the individual species.

Seen from a human perspective, the elephant's behaviour is almost ecological, but how conscious is it? Do they know when they should remove trees, and when to care for them? Is it a genetic, fully automatic mechanism, which varies the behaviour, or is it due to the intellectual handling of outside impressions?

Permit me to insert a story from a completely different world, in order to illustrate how complicated the question of conscious and unconscious behaviour can be.

The best snack an Inuit from Greenland can serve up is *mattak*, a small cube of whale skin with the underlying blubber. This highly prized delicacy will scarcely be to the taste of the uninitiated, but it has nevertheless meant the difference between life and death for the people of the polar regions, since the layer between the skin and the blubber is their only source of Vitamin A. They have thus developed a – probably genetically conditioned – veneration for *mattak* without knowing that it is actually essential for life. Or do they? How can one tell?

The line of development of the savannah elephant has, as already mentioned, split off from that of the forest elephant. It is a new species which has brought new areas into use, or has it created them itself? Did the open savannah exist before the elephant began its forest clearance, or is the open savannah the result thereof? Which came first, the chicken or the egg?

The savannah elephant has probably taken over the Indian elephant's niche, gradually, as the latter disappeared from an Africa which may be supposed to have been less open and more covered by forest and bush savannah than is the case today. Yes, even today the individual places of habitation of the savannah elephant have a tendency to develop dense thickets if they are not kept open by elephants or by fire.

The obvious conclusion to draw seems to be that the development of the savannah elephant has occurred in parallel with the fact that it has created the precondition for its own existence by opening up the savannah a couple of million years ago. Its behaviour and its characteristics have, so to speak, developed through interaction with the current ecological conditions. The more open areas there are with access to fertile grazing, the better the chances of survival and success. This offers a reasonable explanation of why the forest elephant has become the savannah elephant.

It seems as if the spread of both the other grazers and human beings on the savannah took place around the same time as the emergence of the savannah elephant. Our ancestors on the savannah ate gnu several million years ago. The question is, whether the human being and the elephant collaborated in the formation of the landscape. Did they create the great open expanses of the savannah? Today we make use of fire to keep the savannah open if the elephants don't perform the task, but maybe we were already doing so two million years ago.

Perhaps we burned the drought-blasted tufts to lure the prey with fresh grass. Perhaps we burned the savannah to encircle our prey. We had fire, and why should we not have used it rationally? Without fire and elephants the savannah becomes overgrown and we know with absolute certainty that the human being has kept the savannah open the last few thousand years, but if the trail can be followed even farther back we are suddenly faced with another question: what is 'natural' nature? The human being may possibly have created the savannah landscape or been significantly contributory and thus also created the larder and all the preconditions for animals which are prey to exist in the same way as the elephant has created its own 'hunting grounds'.

Sensitive elephants?

An elephant has self-awareness: it knows who it is. It knows and recognises itself. Should a mirror be placed in front of an elephant, the animal would establish very swiftly that it was looking at itself. Draw a mark on an elephant's face and it would rapidly understand that there was something wrong with the reflection, raise its trunk up to the mark to examine and remove it. The elephant shares this characteristic with numerous other animals with highly developed brains. We do not know the extent of these abilities in nature, but provisionally they have also been reported in dolphins, certain apes, pigs, crows and human beings. These animals also understand that the other actions seen in the mirror take place behind themselves and will react accordingly. If something interesting is happening in the mirror image they turn round.

We know that elephants are very attentive towards the sick and the dying. As a rule elephants will try to help sick family members back on their feet. They support and push with their bodies and trunks. Help is only abandoned when it appears hopeless and threatens the existence and health of the others.

Acknowledged elephant researchers describe elephants' overt sorrow over dead family members, a phenomenon that also seems to occur in the other highly developed animals, which lose one of their young, for example. Elephants return at regular intervals to the places where the deceased fellow members of the species lie, lift their bones and sniff them before carefully putting them back. They are complex animals with extensive brain activity: they learn and remember. They can react based on their experiences and predict developments. But which feelings do they share with us?

We humans have defined an ability in ourselves that we call emotion, and specifically under this heading, love; the ability has, into the bargain, many variants, such as mother love, love of one's neighbour (charity) and empathy. Love is regarded as a markedly human characteristic and the evolutionary advantage is quite obvious. The mother's care and love for her offspring and the father's care and love for the family ensure the unbroken sequence of the genes down through the generations.

Love makes the human being go that extra distance, possibly even sacrificing itself in order to preserve the family. This is called altruism and is found in many places in nature.

The survival of animals is something we associate primarily with something instinctive, as with Descartes's pig without a soul, but why this difference? Logically, after all, haven't we developed from the same material and have continued to possess many features in common, which evolution has found advantageous to provide?

If evolution has equipped the human being with an emotion we define as love, what is therefore the argument that precisely this feeling should not exist in elephants or elsewhere in nature? If elephants feel grief, they must also encompass love. One is the result of the other. If one feels grief or love, one has, after all, defined oneself in relation to other existences. One reflects, one thinks. Does one therefore exist?

And then what about intelligence and …

A BRIDGE BETWEEN
SPIRIT AND MATTER

To describe the brain's emotional hemisphere with the rational half is like ripping up grass with a hoe.

What do elephants and narwhals have in common?

August mornings can be bitterly cold on the East African plateau, and the colourful Masai rugs have been put to good use. Open safari vehicles do not offer much cover when the wind howls over the savannah. Even so most enjoy the hour-long drive to the Masai Mara's white rhino, which is kept under constant guard on the periphery of the national park. The bunch of shabbily dressed park wardens around the half-wild animals clashes with the illusion of wild, free, and, not least, untouched nature. It is, however, no more than a fortnight since a leopard forced its way between the rhinos and killed two elands (eland antelopes) which the park wardens had taken in to care for. So this area is not entirely 'tame'!

It is long time since Africa has had that 'free nature', with which we are confronted in the media, and in practice we have to live with ever more intrusive compromises, such as hedges, surveillance, care of nature and rules of behaviour. In fact we ought to be grateful that, despite all, it has been possible to preserve an original impression of how Africa appeared a hundred years ago, if only in fragments.

The rhinos are a gift from South Africa, a present from Nelson Mandela, and presents should be cared for. A previous gift ended up, by all accounts, on daggers in Yemen and aphrodisiacs in the East.

But rhinos flourish when they are not being shot at, and one gave birth to a calf in January 2007, when Kofi Annan visited Kenya to broker the political conflicts in the region. As the calf quickly proved to be a tranquil and level-headed creature, it seemed obvious to name it after the great, world-famous African politician. A few months ago little Kofi's parents were again under attack by poachers who held up the park wardens with Kalashnikovs while attempting to saw off the attribute of the largest sleeping rhino, which, however awoke and resolutely thrust a horn into the side of the presumptuous individual. On that occasion the poacher drew the short straw and ended up in jail severely injured but there is little one can do against well-organised bands, armed to the teeth.

The white, or rather the square-lipped, rhinoceros thrives in many places, especially in South Africa, where most of the reserves or game farms are beginning to breed them. It is said that they do not belong in the Mara – that they have never been native to the region, but who can say that with absolutely certainty? White rhinos can be found both south and west of the Mara, and the conditions for existence in the Mara itself are quite outstanding, so there are in fact no grounds for believing they have never been here.

Unfortunately poaching has become more widespread in recent years. Elephants are poisoned and their tusks removed. There is hardly a month in which police do not confiscate ivory or imprison smugglers. In the Mara we are gradually experiencing a lot of precisely these instances and it is reckoned that some ten elephants perished on that score in the course of last year.

When in 1991 all trafficking in ivory was banned worldwide, the price of ivory dropped by 90 per cent 'overnight'. Previous to this the hunting pressure on elephants was so massive that the animals found their own way into the best-protected national parks, where game wardens offered them some degree of protection. Three months into the ban elephants again began to spread, as if they knew they were safe. Discussion of the ban or the limited sale of controlled ivory blows back and forth. In East Africa, there is the fight for a total ban which is the only measure which has proved to be effective. Controlled sale always results in poaching, because there's a deal of finagling with licences and permits. When at one point there was discussion of quotas, Somalia asked for a quota of 14,000 elephants, but their total population has never been over 7,000 and is probably far, far less. What on earth would they use this quota for?

On the ride back to the camp we scout for elephants. In the last two days the big animals have been few in number. Curious how things can change from day to day. One day we see four or five elephant families in a couple of hours, and the next day there's no sign of any. They must have gone into the elephant house. A promise is a promise and never in my wildest imagination would I dream that we can drive 150 kilometres through the Mara without seeing a single elephant. Not right now. This is however unfortunately the case, and crestfallen, we are returned to the late brunch, where a good round of eggs, bacon and pancakes with syrup must serve as compensation.

The sun has gradually sneaked over the escarpment, and the late-morning coffee is served in its baking rays and Rasmussen's Safari is seen from its best side, when two elderly bull elephants sneak quietly up the steep Oloololo Escarpment to dip their trunks in the Mara. They come out on the drinking place just opposite, where the landscape slopes down towards the river. Long ago all the grass and bushes were worn away by numerous drinking animals.

On both sides of the river there is peaceful drinking, of water and coffee, in appropriate mutual awareness, even if the attention is considerably greater from our side. The elephants stick their tusks into the brink and the hard black earth crumbles. They dig while the trunk searches around between the stumps for delicacies. Small stumps are pulled up and put in their mouths. Do they eat earth? asks someone. No, they eat salt in the surface of the earth, even though they get a great deal of unwanted material in the process. When the water in the earth evaporates, the salt is drawn up to the surface and is most easily accessible here on the brink, where all the grass has been worn away.

The tusks constitute effective digging tools, which are used so diligently that one sometimes breaks. One use of the tusks is chipping away at trees, in order to make the moist inside of the tree- trunk accessible to the smallest of the clan. This is particularly the case with baobab trees, which are significant reservoirs in the dry season. Most of Africa's old baobab trees are crippled after numerous visits from elephants, but new bark grows over the wounds and, generation after generation, the elephant clan visits the same, life-giving tree when necessary.

However the elephant seems to do very well without ivory. Today Africa can boast numerous populations without tusks, or with very small tusks which are almost unusable. The human being has turbo-charged evolution by the extermination of elephants for more than one hundred years. In many areas such as Selous, Manyara and Kruger almost all elephants with tusks have been slaughtered. The hunters have concentrated on the oldest and most powerful elephants with the largest tusks, but gradually, as the giants have been exterminated, the hunt is on for almost any animal except those which are devoid of tusks for genetic reasons. The gene pool without tusks has in some areas therefore become the majority, simply through human selection. Tusks have however had many shapes and numbers, when one looks back at the ancestors, so the question is whether they are a throwback to earlier times, now on their way out, or whether the use of them still constitutes a considerable advantage.

Right in front of our drinking elephants is the herd of hippos lazing around

Cave fruit bats (*Eidolon dupreanum*).

as usual. Once in a while the peace is broken by a loud, not to say disturbing grunting and violent splashing of water. Often this is as far as it goes, but occasionally it is accompanied by the violent, aggressive attacks. It doesn't take much before the hippo puts its teeth into the creature next to it and closer inspection would reveal numerous wounds and scars on most backs. The teeth of the hippo have developed into a lethal weapon, with the long, sharp tusks in the lower jaw and the overgrown teeth in the upper. This violent array of dentistry is completely unsuited for any kind of foraging, where the teeth are simply in the way. This is why the hippo's lips have developed as a substitute for teeth. The lips are so hard that they can be used for grazing, while all the incisors are unused. The hippo's development of tusks may well be connected with the fact that it shares its habitat with the crocodile and in that situation a good set of teeth is not entirely inconvenient.

Nature has a number of examples of the development of tusks or ivory. This includes, for instance, the sea unicorn, the narwhal, as well as the walrus and many breeds of pig. The small characteristic warthog of the savannah has tusks of 20–45 centimetres, fairly massive in relation to the animal's size.

The capacity to develop tusks is genetic, and it is assumed to be the same mechanisms which lie behind the development in various species. All the species mentioned in this context are mammals and have the same common

123

ancestor (as everything has if one goes back far enough), but they have developed along various lines with widely differing appearance, and tusks have turned up once in a while quite independently of each other. The capacity to produce long teeth must therefore be a part of mammals' and thus human beings' genetic potential which – at least in theory – can be developed in all species when outside conditions point in that direction.

Thus a species can develop tusks if there is an advantage, because it already possesses the necessary genetic characteristics. In biology these are called homologous characteristics. A good example of a homologous characteristic is hands, forelimbs, paws and fins. The precursors of hand bones are to be seen first in the lobe-finned fish, the hand-shaped fins of which stayed with them when the fish moved onto the land and developed in many directions with the above-mentioned limbs.

Many of the characteristics we find in nature must, thus, be assumed to potentially exist in the human being and vice versa. All species are in constant development, which means that new characteristics are developed while others are lost, yet what does it mean to say characteristics are lost? Are they in reality preserved in the genome, and can they just pop up again like a jack in a box? One must assume that the distance between a now living species and its ancestor with the lost characteristic is of considerable significance, but nevertheless one of the human being's potential reptilian characteristics, namely the capability of regenerating lost limbs, is being investigated. If it were possible to isolate the reptile's ability to regenerate lost limbs it certainly should not be ruled out that it could be possible to awake the same slumbering mechanism in our own genome. This thought opens up truly extraordinary perspectives.

It should be mentioned that homologous characteristics are still the subject of heated debate because many details remain unknown. The concept is almost a red rag to adherents of Intelligent Design, who naturally must oppose the idea that traits are developed and recycled in new, adapted forms. The whole concept is directly counter to the idea of destiny, but the description of the development of homologous traits in numerous forms is, however, so meaningful and the scientific evidence so overwhelming, not least when genetics is brought in, that I do not want to expend any more energy on the discussion.

Fortunately our two elephants pose for a long time and have amiably allotted plenty of time for the fetching of cameras and video-cameras. They are immortalised together with hippos, half-tame grenadier birds, hibiscus, our tame impala and lovely women in the foreground, while baboons, waterbuck and a single zebra are consigned to the shadowy background. The story of the elephant's tusks has come onto the digital soundtrack so now people are chewing on a piece of toast as well as on my last assertion concerning the elephant's highly disturbing mutual communication about which nobody has heard.

Hmmm! There is clearly also an 'impression' of the elephant's growling on the soundtrack, even if it has to be measured before we have certainty, I permit myself to claim. I can pull yet one more trump from my sleeve and point at a small yellow spot in the thicket: "That thing there makes as much noise as the elephants!" The patch of yellow is a small, yellow-winged bat, which we often see hanging between the thorny branches of the acacia. The insectivorous bat sends out sonar, which is just as powerful as the elephant's at volumes of over 100 decibels. We are unable to hear bats or elephants, and they can't hear each other either, because, fortunately, we all operate on different frequencies.

An aeroplane does not merely fly around in the air haphazardly. It flies in carefully measured-out airspace that we call corridors. It doesn't take much imagination to realise what would happen if the planes began to deviate from their corridors. In the same way the different species each communicate by means of their particular sound corridor. This is the same principle which is used in all forms of radio communication.

If we could hear both the elephant and the bat at the same time, we would hardly be able to enjoy our morning coffee, but sit instead in an infernal spectacle, corresponding to sitting on a runway with jumbos taking off and landing. Even the bat is disturbed by its own noise, and would, theoretically suffer hearing damage without a small evolutionary finesse which closes the auditory canal when it transmits its sonar, and opens it for the echo which returns at a more human level.

An echo from the past

The bat has suddenly become a great deal of the day's story and must be filmed but every time we get close enough to the little super-aeronaut it flutters round and positions itself fifty metres farther away. Fortunately we have plenty of good bat sites, for the unobtrusive little creature can be found everywhere, if one keeps a watchful eye out, in outhouses, under the eaves in trees and in the hollows in hillsides. It comes as a surprise that bats are in fact the most successful family among mammals, and constitute 20–25 per cent of the total number of mammal types, or around 1,000 different species. The numbers can vary slightly from source to source, because neither here is there unconditional consensus as to what should be regarded as an independent species, and what are variants and sub-species.

Once great lizards dominated most of the planet. They had occupied almost all of the vacant niches, and evolution's most recent experiment, mammals, lived in hiding and only crept out under cover of darkness. It is for this reason

Black-backed jackal, with prey.

Topi antelope.

that we human beings have good night vision. When the great lizards died out we took over the daylight, but left the bats in the darkness. There they gradually took over the extensive airspace, where there was adequate room for a new species, as opposed to the daytime when the birds were firmly ensconced in the airspace.

Most of the mammals kept to the ground and occupied all the available niches of the day, where some began to seek food in the rivers and lakes, some learned to swim and others pushed right out into the sea. The night was filled with juicy insects, which would certainly want to avoid being eaten by the birds but the bats brought this paradise to an end. Moths are a tasty delicacy for bats, and perhaps this situation is precisely the reason why we have 'daylight moths'. Butterflies did in fact develop after moths and probably fled from the dangers of the night. The only nocturnal butterfly can moreover, like many moths, hear the click sounds of the bat and has developed an evasive manoeuvre.

We mostly see bats at dusk and during the early evening, when they hunt insects over the river. Later they shift the hunt closer to our naked lights, and once in a while one notices the little beat of wings when these formidable flyers dart past our faces at a distance of 20–30 centimetres without touching each other when as little as a thousandth of a millimetre apart. In the daytime we often see them in competition with the swallows, and in terms of flight it can be hard to tell the two species apart, because of the considerable speed at which they both manoeuvre. Closer inspection, however, reveals the far heavier wing-beat of the bat, which, rather than a swallow, resembles more closely a swan taking off.

Birds and bats have homologous forelimbs, which, like the human hand, can be traced back to lobe-finned fish, but this is also where the homology stops. The birds of our era descend from lizards, which developed feathers more than 100 million years ago. Initially the feathers undoubtedly served as insulation, keeping pace with the lizards' encroachment into cool areas. Later some lizards developed a flying technique, and from these lizards birds evolved which proved better able to deal with the ecological changes that altered the world and wiped out the lizards 65 million years ago. When the lizards disappeared the birds took over all the vacant niches of the flying lizards.

Then, as mentioned, the ancestors of the bats and other mammals emerged from the soil to occupy the vacant niches. The bats did not have the potential for development of feathers lying just under the skin. One cannot, however, rule out the presence of 'feather genes' since the flying lizards developed from the same branch of reptiles as the mammals. The potential of the mammals to develop feathers, lies, if such were the case, so far back in history that it was easier to develop a new solution to the problem of flying.

The flying characteristics of the bat are a 'fairly unique' example of nature cobbling the necessary characteristics together when required. Flying is one of nature's most complex innovations, as is the eye, but nevertheless it was invented in various forms in birds, insects and mammals. Whereas birds have feathers, the bat has stretched skin between its fingers, arm, body and hind limbs. The flight of insects is so complex that we do not actually know how the thin wings developed, but we do, however, have an idea of how large insects such as the bumblebee defy the law of gravity, by clapping the wings together over the body and creating low pressure when the wings slide apart. In this way an updraft is created over and under the wing simultaneously. This corresponds to using cycle shoes which are fastened securely to the pedal. The foot both pushes and pulls.

Completely different, available tools, have led to the same result. The species have thus managed to develop extreme techniques that far outstrip our abilities. All in all this is a solid confirmation of Jacob's comparison, discussed previously, with the *bricoleur*/tinker: if there is a need to fly then a solution will be found. Give me an old tapestry and I'll make a magic carpet.

It is relatively seldom that we have the opportunity to talk about bats on an African safari. They don't cut as grand a figure as elephants, crocodiles and giraffes, but gradually, as it dawns on the participants that there are actually a fair number of them, and that they fly around among the tents, Hollywood takes over. Stories of Count Dracula and vampires are lurking just under the surface. Bats suck blood and transmit dangerous diseases. Can one contract AIDS from bats? Here, however, there are no vampire bats, as the three small vampire species which live by sucking blood are only found in South and Central America, where they certainly can spread such diseases as rabies.

As mentioned, most of the small bats live on insects, even if some also prey on larger creatures, such as birds, other bats, frogs, mice and even fish. Some of the small bats have developed a taste for nectar and fruit, but otherwise the vegetarian habit is mostly associated with the large bats, (megachiroptera) which are also known as flying foxes. That bats enjoy such outstanding success is due, precisely, to the fact that they have shared the food niche among them and limited the interspecific competition: a vacant main niche with many food niches.

A parallel to the vampire bats can be found on the Galapagos Islands, where the vampire finch lives by piercing holes in other birds and sucking small quantities of their blood. All the thirteen or fourteen Galapagos species of finch descend from the same ancestor and in the course of time the original birds have specialised in various food niches and thus developed special characteristics. The vampire finch has developed a sharply pointed beak corresponding to the needle-sharp teeth of the vampire bat. The lust for blood is something the two species have in common with a lot of other creatures. Just think of all the biting insects or of blood pudding. Masses of animals suck the blood of other animals without damaging their prey, which is a pretty smart strategy that permits recycling. Among human beings the Masai have refined the technique to perfection, and today blood from cattle and goats still constitutes their most important source of protein and vitamins.

The bat has a high metabolic rate like its relative the shrew. They have to eat a great deal and cannot manage for too long without food. Acute lack of food is particularly a problem for the vampire bats because, as a rule, a hunt can only have one of two outcomes: a good meal or no meal at all, while other species can have better or worse hunts. For the vampire bat, a couple of bad hunts may mean death by starvation, but fortunately, when in a tight corner, they can get blood from another of their fellow species. Numerous observations in nature

Tawny eagles waiting for the new termite generation to break out of the mound.

and scientific research have actually revealed that vampire bats regurgitate blood for each other when hunger becomes sufficiently imminent and life-threatening.

They are well organised as far as who helps whom and when, so that no one is over-exploited or scrounges without contributing (takes without giving). Research shows that vampire bats only help each other when the lack of food becomes genuinely acute, that they remember who has helped in the past, and that help, moreover, is kept within the family, which should be understood both as the genetic family and other individuals living in the same place. There must, therefore be an evolutionary advantage to this form of reciprocal support and cooperation, which is held together by some form or other of internal communication. You scratch my back, I'll scratch yours. Sounds familiar!

Both Darwin's finches and bats are illustrative examples of how a single species develops and fills niches, even though we do not know all the details involved. Darwin's finches have developed on several very desolate islands over a few million years, where we have a clear picture of the development of the ecology, and bats have developed in an unoccupied niche that is entirely their own, being hunters of insects in the air at night over a period ten times as long. Therefore the diversity and differences are also far greater among bats, but the principles are the same. In the course of a few million years, Darwin's finches have developed into pure seed-eaters, cactus-eaters, insectivores and blood-suckers. The same thing has happened with bats, but the prolonged course of development has created much greater differences and forms of specialisation. Among bats, we find, for instance, nectar-eaters, the tongues of which are longer than their own bodies. There is such enormous variety among bats that at one time it was assumed that they descended from two different ancestors, one produced the large, fruit-eating flying foxes and the smaller, insectivore bats respectively. The problem is, however, that the characteristic traits which apparently distinguish each group are represented in both families.

One of the more peculiar, curious features of the bat is the inverse proportion between the testicles and the brain – a fact at which one of the female safari participants does not seem to be surprised. It appears that males of the species which live in permanent couples develop relatively large brains and correspondingly smaller testicles, while the opposite is the case with species where the females mate with many different males. In the latter case, it is the quantity and quality of the seed which are crucial for whether the single individual manages to pass on its genes.

This may, of course, provide the occasion for considerable speculation concerning the general significance of the brain, where some will conclude that the human being's large brain developed to secure mating rights and that large brains will therefore automatically offer better opportunities in this respect. Even though the connection between the size of the testicles and the frequency of mating is gradually becoming a classic in biology, there seems to be no general correlation between biology, testicle size and brain size. Chimpanzees have large brains and large testicles and mate as often as possible. All bats which make use of echolocalisation of their food need a large, complex brain capacity, so perhaps it is not so much the size of the brain as its structure that we should be looking at.

The characteristics that best ensure survival grant mating rights, and the mating rights that are best employed offer the best survival. A Bushman can catch up with an impala by the use of his legs but not of his brain. The brain may invent a bow and arrow, but neither is this of much use without the muscular capability to use them.

To be or not to be a bat

In the dusk we set out the bat lights. Why they are called that I don't know, but a good theory is that they actually attract bats. Light attracts, first and foremost, all the insects which the bat eats, and thus the bats themselves. Some people get scared by the bats' threatening behaviour, when they fly straight at one only to turn aside at the last moment. Their behaviour is simply due to the fact that their sonar transmits a thin, laser-like sound-wave, directly in front of them, reaching only a few metres ahead. With a flying speed of

50 kilometres per hour the bat has nanoseconds from the moment the echo returns to the ear till it is time to execute any possible deviation manoeuvre. This is the same lapse of time their potential prey has to take evasive action if it hears the signals. There are those which do, and therefore they can still be numbered among the creatures of the Earth.

In a few millionths of a second the bat's brain analyses the sound picture and can, against this background, determine species, size and direction and speed of flight, evasive action and so on and coordinate all its movements consistently with these. The bat is so skilled in these respects that it can fill its stomach several times on a hunt, while simultaneously digesting most of it. The huge consumption of energy actually demands a pulse rate of 600 and a massive metabolic rate. That the bat can do this for more than thirty years without a heart specialist is something of a mystery.

The ability to echolocalise exists to varying extents in many animals, including both birds and mammals. Human beings are also capable of transmitting sounds and decoding the echo from hard surfaces, even if this ability is not especially developed. There are rare examples of blind people developing echolocalisation, which are highly reminiscent of the bat's. They learn to send out click sounds very similar to the Bushmen's tongue click against the palate, and decode the echo. In this way they achieve a feeling of the space around them which helps them navigate their way around. There are numerous other examples of people being able to develop individual senses to a level close to the sublime if they are trained and manage to activate the relevant nerve cells in the brain. Wine-makers and perfume-blenders have developed the sense of smell to a level at which it can earn them a living.

The bat's echolocalisation has been developed and refined through millions of years, but far from all species have used much energy to develop this characteristic. Most plant-eating bats have dropped the echo and manage very well with extremely light-sensitive eyes. They have not developed colour vision but then what use are colours at night? Echolocalisation, on the other hand is so radically developed in the insect-eating bat that it can distinguish prey on a leaf or a tree-trunk, and a fish's fin in the water is more than adequate to establish weight, dimensions, species and speed. We should perhaps also observe that many bats go for larger prey such as mice, birds, frogs and fish. Echolocalisation also provides a picture of immediate surroundings on a level with the best eyesight imaginable and is thus an excellent demonstration that sight and the other senses repose in the brain – not in the eye or the bones of the ear.

The ability to echolocalise has without doubt developed to keep pace with the ability to fly. One can imagine how the ancestors of bats leapt up into the air after insects; perhaps they were good climbers like the apes, leaping from branch to branch and snapping up a moth in the air. Individuals which could 'hang' for a long time in the air fared best. The first leaping bats were dependent on sublime senses of hearing and sight, exactly as can be observed in present-day owls, but the bats learned to exploit their exceptional hearing to pick up the echo of their own sound.

The story of this development is not known in detail, but a comparison with the senses of various now-living species presents a picture of the slow, step-by-step process, because echolocalisation is found in various stages of development, and is put to use in a range of combinations with other characteristics. Thus some species have particularly well-developed sight and use echolocalisation to a lesser degree.

As is well known, some of the ancestors of human beings and bats entered the water and developed into the toothed whales of our own era. This group includes all the small dolphins and porpoises, as well as the larger dolphins like the killer whales and pilot whales and the enormous sperm whales. All have the same outstanding ability to echolocalise like the bat, even though there are great differences in the way the sounds are produced. Various species find their own click, but in the end it is the brain's ability to understand the return signal which is crucial. Can't we after all call echolocalisation a homology and thus acknowledge it as an inheritable potential in a great many species?

Much more research has been conducted into the echolocalisation of whales than of bats. This is particularly because the latter are extremely difficult to track. Dolphins have long been the research field of choice. They seem to take a certain pleasure in human company which we, ungratefully, acknowledge by calling them easy to tame. Dolphins can recognise an object of 2–3 centimetres in size from a distance of seventy metres, and can, for example, distinguish between two metal plates held in front of them where only the back – which they cannot see – is different. They can, quite literally, see into and through other species. Unfamiliar noise, waves and wind do not change the dolphin's sonar image. This is a case of quite outstanding abilities, which greatly exceed our own understanding and competence. Characteristics that demand some formidable processors and hard disks, but can this computer do more than this? It seems undeniably the case, when one examines all the fantastic behaviourist research and exceptional accounts of dolphins' contact with sick or developmentally handicapped children which, in more or less garish versions, force their way forward in columns of newsprint, but that's another story.

Bats and dolphins are just as social as elephants and communicate at least as much. As far as bats are concerned it is possible to state very precisely which few sounds they use to echolocalise. The rest can be used for communication with each other. Some species produce a good ten extra sounds which are put together in various ways and are thus reminiscent of what we call syntax. They have, in other words, a language. What they use this language for is not known to any great extent, much less what other communication forms supplement the language, but it is yet another example of social animals with complex brain structures being tied together in a variable communication system.

Masses of animals have something which resembles language and this is probably also true of birds, but such language is only a small facet of communication which is linked to all the senses. One uses what one has. Communication is not as dependent on a speech gene or vocal cords, as we like believing, just as sight is perhaps not as important as the old philosophers believed – not in the general perspective. Much suggests that speech, like other forms of communication, is the putty which binds the animals' social structures together, just as it does with the human being.

What do elephants, dolphins and bats actually talk about? No way of knowing, unless one actually is one, as the philosopher Thomas Nagel said in his book *What Is It Like To Be a Bat?* This is philosophy's classic old chestnut about consciousness contra natura or matter which again comes into play here.

Descartes was fairly convinced that nature could not move itself, i.e. sense and think. According to his own account he would be utterly astonished if it should prove that in certain limbs it was possible to find those kinds of abilities. The classic discussion deals with how far thought, and thus consciousness, is torn loose from nature, i.e. from the material, or whether all the concepts come from nature and the brain, as the materialist Lenin expressed. We do not know the answer; we can only discuss the question. For Thomas Nagel consciousness is linked to the subjective self. The fact that one knows that one is a bat makes one a bat. A human being knows that he or she is a human being, etc. If the bat has that consciousness it wouldn't care less whether we knew it or not. That's what we think. Are we indifferent as to whether they know that we know or is that in fact the difference? Well, we don't know that either – confused? In any event it's the materialist who has the hardest job arguing for dissimilarity between humans and animals. If spirit can be found in matter, where is the difference? That is, if there is one!

Powerful emotions

On the horizon the sun is on its way down over Oloololo, and we are keeping a lookout for the beautiful silhouette of the blood-red African sunset. A giraffe can indeed also do this, but it is somewhat harder to order into the motif than a static desert date palm.

Twilight can offer great surprises, and it does so today, when a cheetah comes sneaking up towards us over the savannah just a kilometre away. Leaning slightly forward, lurking cautiously, a little late for the big cat, I reckon, which is normally active in the daylight hours but, as the saying goes, never say never, and then it happens: a lightning spurt, zigzag passage, clouds of dust kick up revealing the presence of prey. We cannot see what it has brought down but we get the safari wagon turned round and spurt off ourselves after the scene while I 'entertain' by explaining how incredibly difficult it is to witness a hunt at close range without ruining it.

The cheetah has seized a modest little impala fawn, which is already dead when we arrive. The cheetah's mouth and snout are bloody. It has learned to swallow its meal before more powerful predators, such as hyenas and lions, arrive to steal the prey. We shudder at the tragic sight. The poor savaged baby. The mothers put their hands up to their cheeks or over their mouths. One is, oneself, a mother after all.

After the first camera click, we execute a swift turn through 180 degrees around the scene to keep the sinking sun at our backs. In the process we inadvertently scare a small, new-born Thompson's gazelle, which is lying, well-hidden, in the grass. The baby springs up and zigzags away on its slender legs, revealing itself to every conceivable enemy. The cheetah looks up from its meal and takes only a few seconds to consider before setting off after the new victim. Two for the price of one. It's a great day to be a cheetah. Now there's real terror in the car. Ohhhh nooo! Or something like that. The ferocious predator chases the baby all over the place and those with the coolest heads get some fantastic shots. It develops into a real horror show. The cheetah is totally in control; already full up and with more food in sight. Four, five times the cheetah lets its prey slip away and plays cat and mouse. At one point the small gazelle lies down and regards us with its great eyes, while the cheetah stands opposite it quite peaceably, not looking at all as if it wishes to kill. One hopeful woman thinks they are just playing, as the cheetah is already sated I only just manage a brief contribution before the cheetah takes its prey in its mouth and snaps the slender neck with a quick jerk.

We forget about the sunset and drive back home to the camp with mixed emotions. Some people feel we should have intervened, driven the car between the cheetah and the fawn, since we were a contributory factor in the killing, while others believe it was just the way of nature. I am, as always, ready with a little salt to rub in the wound, for they are halcyon days for the predators when the antelopes of the savannah all bring their young into the world at the same time. There are many long periods with edible fawns and without these the cheetah would live a miserable existence as the worst and weakest hunter among the big cats.

It's not exactly rare for safari participants to want to intervene when sweet little creatures enter the danger zone. Can't we do something or other? Naturally we never do, even though, on the other side, we do everything to avoid helping the predators which have learned to camouflage themselves in

Slender mongoose at elephant dung.

the shadow of the safari vehicle. Exhilarating, dramatic or just a little tragic are the words used most often to describe the scenario. Most people seem to believe that nature is cruel, but that is really nature's way. Nature's cruel unlike human beings? I ask, a touch provocatively. It doesn't take much imagination to envisage what sort of openings this can lead to, but it remains an interesting point that we do not really know what is actually happening. Why do we feel the urge to help young animals in total opposition to the so-called 'way of nature'? No one is in any doubt that a lot is going on in the head, and that it can be related to something we human beings call emotions. Without feelings the experience would probably not be interesting, but the experience is a two-edged sword where we are both attracted and repelled by the horror. Isn't it the rational mind that bestows its seal of approval while feelings are revolted? Is there really a difference at all?

Back in the camp we meet John, living proof that a cook from the Danish island of Ærø can become a hotel manager as far out in the African bush as it's possible to get. He is also living proof that the way of nature is not always of prime importance. Like his patrons he cannot be responsible for the small, young animals of the savannah, and certainly not for abandoned young antelope with great doe eyes. Often we have a menagerie of antelope young feeding from the bottle which have to be tethered outside the office so that they do not get under the feet of their new mothers. We also have other things to do.

The guests are fond of our pets, and so is the leopard. It loves children and dogs of every kind and slinks around in the dark night summoned by youthful sounds which it identifies from a long way off. It is no simple matter to look after the little ones, which we kindly lock in every night – that is if we can find them. Once in a while they feel compelled to hide in the bushes, a move which, very probably, will lead to an early end to a career, to the considerable sorrow for all the carers and patrons. The sorrow is especially deep when we find the remains. One day we attempted to release an impala some fifty kilometres out in the bush, but a day later, exhausted and starving it banged on the door to the office, wanting its bottle. Great joy in recognition on all sides, even if trouble beckons.

Some years ago a film crew became chance witnesses to an utterly peculiar rescue action in Africa in which a crocodile was well involved in dragging an antelope fawn out into the water. Straightway a female hippo came storming along making an instant frontal attack on the crocodile, which released the prey and slipped back into the water hole. Then, with the utmost care, the hippo took the fawn in its mouth and carried it in safety farther up the bank, a fair distance from the still-lurking crocodile. Here she carefully put down the antelope and tried repeatedly to lift it up on its feet by putting its mouth in under the antelope's head. In time it dawned on the hippo that there was no life left in the little animal and after a couple of feints toward the crocodile, it turned and left the place. At last the crocodile could fetch its lost prey. Without the requisite film documentation the story would hardly have been accepted in the more scientific world, where the episode would naturally be explained as awakened maternal instincts. Scaring and attacking crocodiles is everyday fare for hippos when they have young. It is possible that the hippo in question had lost a calf and with carer hormones coursing through the blood, everything is possible. Who knows?

It is, however, remarkable that, in the same instant that hippopotami become the subject of conversation, we change our mode of speech from emotions to instincts, quite in the spirit of Descartes. When we use the word 'instinct' about an animal's behaviour, it is, as a rule, to put a distance between animals and mankind. We downgrade the animal's patterns of reaction to something instinctively innate, which is not linked to brain activity or considerations. Animals have no free will.

Elephants, apes, dolphins and many other animals display the same devotedness to their own young and close relatives as human beings, and something suggests that a certain devotedness can be stretched to more than one's own progeny (genes). Is it the same mechanism that makes us give the feeding bottle to young animals and shudder when they are killed? Are the reactions of the hippo and ourselves exactly the same, and prompted by the same mechanisms? Is it the brain's reward proteins or neurotransmitters that are at stake here? Are these feelings called up alone by these valuable brain chemicals which obviously make us feel good? And do the many different events parallelise and evoke uniform emotional reactions?

Is safari for dopamine junkies? There are certainly many people who would not take too kindly to the thought, the idea that feelings are chemistry. But why should this make them something inferior? It is, after all, the thought that counts.

Some weighty tomes have been written about powerful emotions and more scientific papers about the anatomy of the emotions. We can even exert ourselves to dissecting emotions and categorising them for various opportunities and situations, even as precisely as, occasionally, we have borrowed the prestige-filled word 'intelligence' and mated it with 'emotional' to give us emotional intelligence, social intelligence etc. That's absolutely certain!

When help is self-help

Like most of the savannah regions around the world the topography of the Masai Mara is marked by small eminences of earth which undulate across the landscape like a restless sea. These are the work of small termites, which, as it happens, are more closely related to cockroaches than to ants or bees. The termites are ant-like in their diligence, and their redistribution of the soil creates shelter, shade, concealment and watch-towers to the joy of many creatures.

One of the most characteristic sights of the Mara is a solitary topi antelope stationed on top of a termitery surrounded by various peacefully grazing antelope and zebras. Thus members of the same species and other grazers make use of the topi's lookout service and have learned to decode its reaction patterns. When the topi get going everybody gets going. Lookout systems are also known to exist among birds feeding in flocks, where there will always be one glancing around for possible danger. The topi has probably developed this behaviour to the delight of his own family and species, and subsequently others have got into the act, just as many mammals have learned the bird's warning signals. Cooperation with nature seems to function across family and even species, and gradually, as we discover more examples, it gives rise to considerable speculation, for what mechanisms direct these relations?

From a Darwinian point of view the behaviour of every organism is, after all, controlled by the drive to pass on as many genes as possible to the next generation, and one must therefore interpret all behavioural reactions in this light. This rational interpretation of nature means that selfishness is 100 per cent and that all actions should be interpreted with mathematical logic. A leads to B when the preconditions are fulfilled. B is predictable. The theory will thus be challenged by actions which do not seem to increase an animal's survival skills or its ability to transmit its genes onward.

At first glance it costs the topi nothing to act as watchdog, because it can stand and chew the cud in the meantime and await the emptying of its stomach. On the other hand, it succeeds in establishing a protective wall around it, and masses of other options for a predator to attack. One of the most important strategies which I described earlier is precisely being one of many. There is a big difference in being one of ten or one of a thousand. Everybody wins by this arrangement, though a small problem needs to be solved: there must be a regular changing of the guard, so the one on watch can eat and rest as much as the others, or else its chances are substantially reduced. We must assume that the topis have this situation under control and change positions at regular intervals.

Another of the savannah's antelope species goes further when it's a matter of helping each other. They are the impalas, which remove each other's parasites from those places difficult for the host to reach, such as the head, the throat and the neck. The arrangement is a mutually practical and not determined by the bonds of kinship. With many other animals, such as giraffe, rhino and hippopotamus the parasite problem is solved by a small bird, the oxpecker which removes up to 400 parasites daily, but the impala has developed another and somewhat curious system, which gives behaviourists grey hairs. When they remove parasites from other animals they use energy and time which could be expended on feeding. If everybody helps each other equally then the money goes round properly when the energy bill is totted up. If one individual cheats on the balance scale and scrounges off the company, then an imbalance occurs. The scrounger will have a better chance at survival and the scrounger clan will grow and gradually take over at the expense of the 'good guys' that, on their side, need to make themselves safe from scroungers.

We may conclude that there must be some form of balance or else the system will simply not exist. This is not a matter of altruism since nobody has an immediate interest in sacrificing oneself for others when this is viewed from a genetic viewpoint. All in all it really does remind one of something we recognize from the society around us. What mechanism is actually behind this reciprocity and the fair division of 'goods' is something about which we do not know so much, but there is clearly a great deal of food for thought.

Nature is filled with examples of parents sacrificing themselves in the fight for their progeny's survival. It is thus quite common that birds use themselves as bait to distract predators from the nest, just as baboons sacrifice themselves to an attacking leopard in order to protect the clan, or a zebra stallion forms a front against an attacking pride of lions, while the females and foals remove themselves to safety. Consistent with this is protecting one's own genes, there is also a tendency to support genes with more distant relations (e.g. cousins). This can be seen, for instance with lions, where the young are suckled by

A Lilac-breasted roller atop a wildebeest (gnu).

all the females in the pride which, as a rule, are closely related. With lions, however, there is a clear correlation between the quantity of milk given and the closeness in terms of genes. The mothers always give their own cubs most milk. When the female gives milk to a nephew, niece or grandchild, with whom she shares 25 per cent of the genes, this must not happen at the cost of her own cubs' chances of survival (fitness). It is, therefore, possible to conclude that the system ensures her own young a considerable average chance of survival at the same time that the survival of the related gene pools is ensured (little is better than nothing). Reciprocity is help to one's own genes.

In recent years interest has grown around reciprocal social behaviour, known as mutual or reciprocal altruism, which appears to take place between non-genetically related individuals. Ordinary altruism occurs when it is an advantage to one's own genes, completely in accord with the theory of evolution. But what reasons and mechanisms control the reciprocity when it is not kinship? How widespread is this behaviour? Which mechanisms control it? How do animals experience reciprocity? Are emotions involved?

We are talking, therefore, of behaviour where an individual apparently uses energy to the advantage of another individual to which it is not related. Among human beings this would be termed love of one's neighbour, social adjustment, helpfulness, etc.; type of behaviour which is completely inconsistent with evolutionary biology, because this behaviour would simply be extirpated through natural selection. And yet it exists among human beings?

The phenomenon of reciprocal altruism can be pared down to "If you scratch my back for five minutes, I'll scratch yours for five minutes." Thus we have the connecting link to Darwinism. All actions are, at bottom, selfish. The impala cleans its colleague's pelt in expectation of receiving the same treatment itself. Like for like: 'millimetre democracy' might be a useful label. We know that chimpanzees help each other within the clan, and that individuals which attempt to wriggle out of reciprocal obligations get punished. Yet we still lack the answer as to how one controls this system of quid-pro-quo arrangements, and, perhaps more interestingly, to the question of how far the rest of nature is different from mankind, and whether there even exists true, self-sacrificing behaviour. Do individual people sacrifice themselves for the whole group? And is it not, in the final analysis, a question of one's own gene pool?

A hairy aeronaut with emotions

In order to get a spade-thrust deeper down, we must once more turn to the bats which probably display the best-explored example of mutual or reciprocal altruism. These are the vampire bats, which, as already mentioned, live by sucking blood from other animals. They fly out of their holes at night to forage. When a victim has been echolocalised, the specialised, heat-sensitive cells help the bats find their way to the surface blood vessels. Needle-sharp teeth bite through the skin with a refined, acquired technique to which the victim pays no attention. An enzyme in the spittle prevents the blood from coagulating and the bat is ready to fill its belly. A meal takes twenty to thirty minutes and every night they have to take in the equivalent of 50–100 per cent of their own body weight to avoid starvation. Two nights without food is life threatening, and after sixty hours without, they lose up to 25 per cent of their body weight and can no longer maintain the body temperature necessary to sustain life.

It is well documented that deficiency in hunting success strikes at random and is not due to deficient hunting skills except among the very young. The risk of an unsuccessful hunt is there for the animal's entire life. One third of the youngest must return with nothing to show, but gradually as biting technique improves, the number of unsuccessful hunts falls to around 7 per cent. The quick, energy-rich meals are an evolutionary advantage, but there is also another side to the coin. The enormous risk of losing a meal has necessitated an insurance system. It is actually, as previously mentioned, possible for the starving bat to feed from some of the members of its own group, who have had more success on the night in question. They regurgitate some of the contents of their stomachs into the mouth of their begging fellow members of the species, and thus increase the recipient's chances of survival. As one might expect, the greater part of this kind of allocation of food is between mothers and their pups (as modern specialist literature refers to young bats). Already at a very early point in the pup's life the female supplements mother's milk with regurgitated blood and this behaviour continues after the pup is grown, until it has learned how to hunt properly.

There are only a few scientific studies but one survey documented that in 30 per cent of the cases when adult bats donate blood, it is to unrelated females and their pups. Even though the recipient is not related to the donor, these are not random occurrences. It is a case of individuals the donors know well from the roost, and with whom they have shared nesting locations, possibly for many, many years. Some females stay together for years (even if they're not related) and tend to exchange blood mostly with each other. Friends for life, indeed.

Even though mutualism primarily occurs within the group of mothers and young, there are also examples of males regurgitating blood for other males, females and pups, and that females also, in rare cases, give a portion of blood to males of the same community. There are likewise examples of females adopting orphaned young. The system can thus function across boundaries of sex, generations and genes.

When animals share food, it looks like altruism where the donor sacrifices its own food that would otherwise have ensured its own survival. But it is not quite that simple: with blood donation the recipient's chances for survival increase, without the donor itself suffering any loss, because the donating only occurs when the recipient has no chances left, and the donor still has maximum time in which to find food. The accounting is very simple. The bat can, as mentioned above, survive sixty hours on an empty stomach. When it returns home from a successful hunt, it is completely 'full up'. It has, in other words, satisfied its own needs in the hunt and can, in principle, easily get through a whole day and a night. But if it is already hunting again after twelve hours it can give half of its food to another. The two of them each still have food for twelve hours in their stomachs, before the next period of hunting, when the countdown from sixty hours begins. The donor's chances are not reduced, and the recipient has been given another chance. The donor has thus also got something to its advantage as the colleague is now in debt. The donor can depend on help from the recipient, when and if this should prove necessary one day.

The system works because bats can recognise each other. According to a very recent theory, social grooming is an important factor in recognition. Hungry bats often groom the fur of potential donors. This may be a request for food. Another possible mode of recognition is the individual call. The system should exclude freeloaders.

It looks as though the mutualism experienced among bats is not static but that the bonds between individuals are dynamic and can be understood in relation to the prevailing ecological conditions, so that both the males and females are genuinely in a position to help each other when this is necessary. For now, anyhow, we have only indications. Relations must be built up over time: the longer the time, the closer the social contact, and thus the greater the mutualism. With both bats and apes a great deal of time is spent grooming each other's coat; there are friendly hugs, resembling the familiarity in evidence among elephants and whales and many other creatures.

The exchange of food and mutual assistance is always accompanied by previously established close contact that is grooming, not just immediately before the action but also after a previously established build-up of social relations. When and how fast can social relations be built up? And in what circumstances? The results so far indicate that there may be more complex relations, in which more senses are involved. As emphasised earlier, bats, like elephants, have highly complex sound images, reminiscent of syntax, which are reserved for mutual intercommunication. Social contacts are believed to activate the senses of touch, sight and hearing, but as of now any limits to the possibilities are imposed by ourselves.

What can one call the social bonds which are formed between animals if it is not, pure and simple, maternal instinct? If unrelated animals build up friendship-like relations over time, what is one to call whatever it is which binds them together? It can't be instinct! It is probably brain chemistry which, to the point of interchangeability, resembles ...

The study of vampire bats and other social animals probably brings us closer to the actual altruism which is believed to exist in the human being, where the individual gives up something to the whole (society) without necessarily getting absolute value for their money.

Herbert A. Simon suggested in an essay that there may exist a genuine altruism under which the individual human being accepts a higher purpose and relinquishes some of his absolute capacity for survival (fitness) – though on the condition that the individual obtains a certain, guaranteed, net fitness in return. Society offers a guarantee that things cannot go completely wrong, and in return one must abstain from breeding and raising 500 sets of genes like some latter-day caliph. One might call this model anti-reciprocal. The surplus, so to speak, is contractually paid to the group, which should be an advantage for the total population. Society or Chaos.

Transferred to the vampire bat this means that a similar 'contract' among them counteracts the average lack of hunting success. The problem strikes all equally and all have, therefore, an equal interest in solving the problem. Simon's suggestion implies, however, that not all receive the same disbursement, and that in time, therefore, their lines will die out. Those which receive least do not have the same chances, irrespective of how one looks at it. If the whole population benefits, the fitness of some will be forfeit, and there will be shifts in fitness.

A perfect balance can hardly exist, the thought clashes with the dynamic of difference, which is precisely what constitutes the driving force of evolution.

Both cheats and friendly, self-sacrificing souls are part of the order of the day among human beings and animals. There will be periods when cheats and con artists manage to take control in a population, which may perish, or perhaps other factors will neutralise the discrepancy. Should anyone observe how much force the human being uses to neutralise differences and frustrate fraud it is quite in order if a hairy little aeronaut has solved this complex problem.

Griraffes at Entabeni with the unmistakable Hanglip Mountain as a backdrop.

Tolerance is a useful quality in plants and animals; which apparently vanishes when they are civilised.

Africa's island kingdoms

The hole in the great baobab is absolutely circular, and one can see right across through it. Someone or other could have kicked a football straight through the thick trunk. We each stick our head into the hole, have ourselves photographed 'inside the tree'. We experiment with camera angles, and see what we can get into range of from both sides. This succeeds in immortalising one of the Ruaha National Park's other baobab trees in the 'belly' of the great tree. In addition we have set our hearts on photographing an elephant framed by the hole. A symbolic image of the damage and its cause; but elephants do not pose to order.

Regardless of where one looks, the baobab's curious naked boughs are seen against the blue sky, elevated far above the scrubby bushes of the savannah. Now two foraging elephant families are also approaching from the horizon, though still out of range. We can wait. The ancient baobab trees are severely deformed. The trunks are full of large lumps, where the bark has grown over old holes and fissures. For generations the elephants have visited the same trees in order to dig out the wet in the dry season. The baobab is, in fact, the world's largest water-containing plant. The trees bear the marks of the elephants' tusks, but regenerate at an astonishing rate as a result of thousands of years of evolutionary adaptation.

The world is changing however, and

Africa's elephants have gradually less and less space as civilisation takes over 'free nature'. Nature has been transformed into islands in the cultural landscape, which means new adaptations and new developments. To what degree the elephants' relationship to the baobab tree will remain intact, only time will tell.

We wait for the posing elephants and talk returns to the distinctive trees. Why are the trees all alone among acacia and low bushes? Why have they not developed ten, twenty or even more species? Well, off the cuff I couldn't really say, but then there's no law against making use of what you know. Like all organisms the baobab has its own strategy and fills its own niche, which can be more or less distinctive. This particular species is capable of amassing a great quantity of surplus water when it rains and storing it for the hard times. Water, which otherwise runs away or evaporates. It is, thus, water which will not be of immediate benefit to the near competitors. In the same way that the giraffe has its own food niche, the baobab has its own water niche. Not a bad characteristic in an area short of water. When the tree grows above the average height for the savannah it gains more light and can be content with a shorter growing and reproduction period with leaves and flowers. In this way the water consumption is concentrated, and the naked tree gives the minimum shade over other growths. The baobab is a courteous tree which lets others come in for a share of the spoils. There is both light and water for non-competing species, but in this manner they also gain a protective barrier against possible competitors that want a place in the same niche. There isn't room. A corresponding, unique story of adaptation can be told about the only little thistle which found a splendid spot between the great tree's roots.

Our baobab, however, is not utterly deprived of kinfolk: the planet actually has a whole seven other species, which all resemble their African cousin: six in Madagascar and one in Australia. The original tree developed in Africa, or rather, in the primeval continent, Gondwanaland, of which Africa, Madagascar and Australia were a part before they separated. In new surroundings the tree initially developed in three directions. Why Madagascar's one species became six, while two other areas did not develop more species is anybody's guess. The most logical explanation is that the smaller landmass which constitutes Madagascar has had far fewer species than the two great continents, Africa and Australia. The existing species have thus had greater opportunity to spread into new niches, gradually, as the climate has changed.

This development can be seen on all isolated islands, and therefore biologists have been deeply interested in islands since Darwin and Wallace described the development of species in the Galapagos and the Malaysian archipelagos, respectively. Today we talk about islands in a biological sense, when a given natural region is isolated from its surroundings by, for instance, water, ice, fields and towns.

In Madagascar the seeds of the baobab, like the seeds of other plants, possess a certain number of genetic variants, mutations. Once in a while a genetic variant falls on an inappropriate spot, but unlike the other seeds, the variant manages to survive on a lower quantity of water, a greater amount of salt or less light, and thrive brilliantly because nothing else does. This is how we imagine the development of a new species being initiated. The process may repeat itself many times and create many new modifications. This is what we call evolution. Exactly the same thing happens in Australia and Africa, but here, as we have seen, no variant has managed to gain a foothold, possibly because there is not that much room – vacant niches – or perhaps the seeds have not been so 'lucky'.

The elephants are not after the same things as us, so we must content ourselves with pictures of the baobab trees, which are now also part of the sunset. Just a moment, I must have said island once too often, right? I used it about the national park in the cultural landscape and about the biological islands. Ruaha is fortunately – like many nature parks in Africa – not an island in the biological sense, because there is massive exchange with its surroundings. The elephants and other animals can wander freely out and in, but that won't last for ever.

Madagascar and the Galapagos, on the other hand, are islands in the biological sense, because the world's oceans isolate them from the rest of the planet's landmasses. Paradoxically, however, it seems as though Madagascar has functioned as a springboard for species on their way to the Galapagos, and that some species have actually taken the long route across the ocean, while others have made the journey round South America. Turtles and tortoises, insects, plants, birds and bats have found their way to the Galapagos over the sea. However only a very few species actually managed to make the trip, but, they have, on the other hand, spread into a number of niches in the same manner as the baobab trees of Madagascar.

The large elephant turtle came floating across the sea, possibly from Madagascar, even though it's actually particularly land-based. On the Galapagos Islands the tortoise, or turtle if you prefer, developed into a long series of specialised species, each living quite separately from the other. First however, the barren islands underwent a long development. Movement in the Earth's crust pushed them away from the active volcanic zone, where they were formed, the volcanoes had to cool down and become less active and the volcanic basalt had to weather and form layers of earth mixed with washed-up seaweed and algae. Finally, birds, wind and ocean currents had to bring with them seeds capable of germinating. Here the pioneers of life had to start all over again, and great numbers of them had to perish in the process. Today it is still possible to see the hardy lava cactus taking over the first open fissures in the basalt rock.

The spiny palm forest in the Seychelles.

The famous coco de mer palm.

The islands' two extant species of bat represent the family's oldest, insectivorous branch and have hardly played a major role as dispersers of plants. Like many of the other species they have used a more dry-shod corridor via the Far East and South America. Thus not a great many plants and seeds. Therefore the first shy immigrants, such as the daisy, achieved outstanding success once they took firm root. They spread into numerous niches and over a few million years developed into at least fifteen utterly different species and variants. The daisy of the Galapagos today is not only a small flower one can pick with the hand, it has also developed into large sturdy bushes and tall trees (*Scarlesia* spp). A textbook example of how flowers can develop into trees in the right conditions: that it is impossible to intuit anything at all concerning a species' potential for development, merely by looking at it.

In the same way small, fertile massifs with dense forest project from the cultural landscape of Central Africa. Small islands, completely surrounded by cultivated land. Here insects, amphibians, reptiles, flowers, trees and a few bird species live completely cut off from the surrounding world. They have no exchange with fellow species outside the island and develop distinctive characteristics which are adapted to their special habitat. Today, generally speaking, every expedition to these massifs brings to light new species.

Flowers and bats

Ruaha is just one of East Africa's many unique national parks that is not greatly visited, as it lies some way off the beaten track compared to the current safari routes: so much greater the experiences when one can keep them to oneself. The inaccessible natural areas can, once in a while, offer specialities such as gorillas and chimpanzees, but, as a rule, the greatest attraction is actually the peace and quiet. This has both a positive and a negative aspect: one should after all remember that the revenue from visitors ensures the existence of Africa's free nature. No money – no nature. In this way I can reassure safari participants who think they have sneaked into a forbidden area. Enjoy it while you can – you're paying for it, and, yes, you are indeed fantastically privileged, but there is another side to things.

Following yesterday's elephant-baobab safari, I've set a little task while we relax by the River Ruaha in the hottest hours of midday, as one herd of elephants files past after another. Brains get twisted around trying to think of the name of the small, newly planted tree of which I stubbornly claim we have seen so many. I can guarantee that everybody knows the name. There are loads of suggestions – I give in and reveal the mystery name: baobab. The solution is by no means easy to arrive at since the young trees look quite different from the adult ones, the salient characteristics of which are the five lobes on the leaves which may remind one of the chestnut tree. The small plants look more ordinary with entire leaves. The transformation to a 'real' baobab takes place many years later, and therefore many African people believed that a god had put the tree on Earth exactly as it was.

Most flowers open in the daytime, when wonderful aromas and gleaming colours entice the pollinators, but in this respect too the baobab tree has discovered its own path. Like the bats it has turned its back on the day and only blooms at night. When darkness has fallen, the big white flowers open, ready for pollination and the following morning, now curled up, they fall to the ground. The whole process has to be managed on one and the same night. The flower smells neither of roses nor violets, but rather of the dung-heap, or putrefaction, and this rank smell lures carrion flies and the other nocturnal insects. The flies, however, play a fairly subordinate role as pollinators, since the tree's existence and development is safeguarded by large fruit bats which flap around in the corollas, where their fur is literally coated in pollen dust, while they nip the fresh nectar and carry the pollen from flower to flower.

Baobab and bat have a long, parallel history of development, which probably goes back 40–50 million years. Not even here do we know the details, and must have recourse to Darwin and logic, until genetics itself reveals more. The present result looks pretty impressive, for the tree seems to meet the bat's needs. It blooms at night, when the bats are active, with large white flowers, which are easy to see, and the tree's enormous production of nectar is sufficient to cover the bat's massive fuel consumption. The flying fox, as some call the bat, has, on its side, developed an extremely long tongue and a lightning-fast contact technique as bats cannot remain motionless in the air like a humming bird or sunbird. Finally one must remember that the bat has also developed a taste for vegetarian meals. One more, and

Male coco de mer, with a gecko.

Gecko on a palm leaf.

Top: Frog in the tropical rainforest where the humidity allows it to stay away from open water.

Above: Millipede in a tree. They are detritivores and play an important role in comsuming rotting vegetation and fungi.

in no sense an insignificant, adaptation.

There is of course an explanation for the rotten stench of the baobab flowers and the attraction of many small insects, which leads to more guesswork. We do know that the developmental history of many species of bats goes back 50–57 million years, and that in all probability they are related to such insectivores as the shrew. Everything suggests, therefore, that they began their career as nocturnal insect hunters. The baobab tree was originally pollinated by insects which were lured by the flowers' putrid stench. It is, thus, probable that the first bats searched for insects in the baobab tree's flowers, acquiring, in the course of time, a taste for the delicious nectar and having slowly modified their diet at the same rate as their physiology has altered.

Today the interconnection is so prominent that one can lay two distribution maps of the baobab trees and the straw-coloured fruit bats over each other, and state that they fit together like hand in glove. The baobab and the fruit bat safeguard each other's existence. The baobab trees of Madagascar are pollinated in exactly the same way, but by another bat, the closest African ancestors of which still eat insects. This theory rests therefore on a very firm foundation, but as so often in evolutionary biology we have to content ourselves with reason without having any real, solid, scientific proof. The stories of the principles of evolution repeat themselves all over the place if one cares to look so let us just take a glance at a parallel example of adaptation.

Let us take a leap for a moment to the other side of the Earth, to the Sonora Desert in the frontier country between Mexico and the USA on the Pacific side, in order to examine the connections between the world's largest cactus and bats. The enormous columnar cactus, the cardón, grows to more than twenty metres in height and weighs up to twenty-five tons. Like the baobab tree it opens its flowers at sunset and closes them again in the morning. Here there is also a bat involved. History repeats itself. Both the baobab and the cactus are enormous water-filled growths which dominate their respective environments. Each has developed in its particular niche. The baobab tree is indigenous to the tropical savannah regions with abundant seasonal rain, while the cardón cactus is native to one of the world's driest, hottest desert regions, but the fruit bat is something they have in common.

The great cactus has made its flowers easily accessible, high up on the column and, like the baobab tree, produces the large white flowers which are highly visible at night. Not until the evening, when the sun has gone down does the cactus begin the production of nectar which is at its maximum several hours later, when the bats begin to swarm around them thrusting their heads into the corolla and shooting their long tongues right into the nectar. A split second later and it moves on to the next flower.

The small, long-nosed fruit bats have adapted their entire life-cycle to the cardón cactus. They come from the north, from Arizona when the cactus blooms and they savour the nectar for several months. When the plant ceases flowering they move farther south, leaving the fruit and seeds to grow in peace. Before the ripe fruit fall, the bats return to eat them and disperse the seeds. Cactus seeds germinate better when they have undergone a trip through the bat's stomach, and the strong acid has partly broken down the seeds' hard shells. This is a phenomenon which is known from numerous other species worldwide, where seeds often thrive better after a journey through the animal's intestines. Stomach acid also sorts out all the unfit seeds and one can imagine that the evolutionary process can proceed relatively fast if most of the fruit and seeds get eaten before germination over a longer period.

The seeds are spread in large numbers and the few which fall down into the desert's sporadic low bushes and end up on a shaded place have a chance of becoming a new, lush cactus in these extreme surroundings.

Even though the interaction of the baobab tree and the cactus with the bat has the character of repetition, the development is staggered in the two places. The American bat has brought with it many of its own characteristics from African ancestors and built further on them, so that today they have adapted to new plants.

The Paradise Islands of the Indian Ocean

It happens that we hop on board the plane from Nairobi to Victoria, the small capital of the Seychelles on the island of Mahé. Most people regard the trip to the Seychelles as a relaxing conclusion to a glorious East African safari.

However the islands can offer masses of 'nature' though their exciting stories are hidden deep in the lush forests.

The group of islands constituting the Seychelles lies on the equator north of Madagascar. While the islands of the world's oceans are usually formed by volcanic activity and coral reefs, the Seychelles are large lumps of East African granite, which fell off when a substantial part of the Earth's crust tore itself loose from Africa and drifted eastward where it became today's India. For this reason the Seychelles are some of the world's oldest and most isolated islands, far older than the Galapagos.

In spite of the persistent attempts of seafarers and traders to destroy as much as possible, the islands continue to be a living museum with numerous ancient stories of adaptation waiting to be discovered. The greater part of the Seychelles' animal and plant life only exists here and is therefore endemic. The various species have simply been here so long that they have developed their own lines, which deviate sharply from the original species and from sister species on the African continent.

A single type of fruit bat has found its way here and is responsible for the pollination of many different trees and bushes. It belongs to a large genus with more than fifty species (*Pteropus*) which have spread throughout the Indian Ocean to masses of islands, to India, Indonesia and farther south to Australia as well as northward to Japan. Each of these completely or partially isolated species has developed with its own behaviour and biology. Their closest relative is to be found on the Aldabra Atoll, a thousand kilometres away. Both species are more distantly related to species on Madagascar, the Comoro Islands and the mainland.

The landscape is protected and, unlike the nature reserves of the mainland, invites rambling and independent study without the risk of being eaten by wild animals or bitten by poisonous snakes. That said, however, most people will have a serious need to have explained to them what to see and where to see it.

One early morning we are wandering through the national park, Valle de Mai on Praslin, where the important sight bar none is the rare coco de mer palm. The palm produces the world's largest seed, weighing in at 20kg, which is famous on every coast of the ocean where they have occasionally turned up and have been incorporated into folk medicine without people having the least idea where they come from. Their shape reminds one of a woman's lower body, which in itself indicates excellent qualities, and the over-dimensioned seed has also attained a myth-like status. The nuts were soon sought after in Renaissance Europe, but a particularly enterprising captain filled his ship with nuts and set fire to the trees in hopes of increasing the value of the cargo. That put a stop to the export, but fortunately the species survived together with numerous other rarities.

We have engaged a local nature guide who is highly skilled at finding everything we wish to see. We have hardly set foot on the marked-out paths before we see the first 'rarities': the blue pigeon and the black parrot. Soon we also catch sight of the green gecko and the yellow-eyed gecko which both pose for the camera. The atmosphere in the forest is heavy, hot and humid, utterly different from the pleasant coolness of the savannah.

I notice that all the young shoots of the endemic palms (for con-

Giant tortoise from Aldabra, Seychelles ... the last frontiers.

noisseurs: *Nephrosperma vanhoutteanum, Deckenia nobilis, Phoenicophorum borsigianum, Verschaffeltia splendida*) are filled with needle-sharp thorns which apparently vanish gradually as the palm grows and are replaced by stems. In a mildly teasing fashion I ask our eminently capable nature guide if she can explain the sharp thorns. She only shakes her head as if I'd asked a rather silly question. Plants have thorns because they have thorns! There is always a rational explanation; it's just not certain whether we know it or are in a position to put forward a workable hypothesis. I suggest that this is the plants' way of protecting themselves against tortoises. Fair enough, but there aren't any tortoises around anymore, are there? And, of course, in this she is correct, because the European seafarers boiled them all up for soup several hundred years ago. However, before that time all the islands were filled with enormous land tortoises that were related to the Galapagos tortoises, which we still find in tremendous numbers on the aforementioned Aldabra Atoll.

The palm trees on the Seychelles are found nowhere else. Young palms and other plants do not form sharp thorns without good reason. They are a protection against something or other and here there have never been any dangers more serious than the now-departed tortoises. In time the thorns will disap-

Top: Green-backed heron. Found in mangrove creeks and a common resident on African shores, and farther inland.

Above: Electric ray (*Torpdinidae*) found along the Indian Ocean coastline. Do not touch.

pear, but it will take many generations. Yet another example of a highly plausible explanation without any scientific evidence, but let us move the tortoise dynamic to the Galapagos.

On the Galapagos the giant tortoises developed a taste for another plant in the absence of the young palm shoots. This is the hardy cactus fig (*Opuntia* spp.) which is thought to be one of the islands' early pioneers together with the daisy *Bella perennis*. The tortoises have not just developed a fancy for the sweet fruit of the cactus but also evince great veneration for most of the juicy, sap-filled plants, the thorns of which do not seem to offer any serious protection from the tortoises. On islands without tortoises the plants are prostrate, but on islands with tortoises they have developed a stemmed form. Tortoises and cacti have fought a long evolutionary battle that cannot be differentiated from predator–prey development.

At the bottom of the stem the stemmed forms have developed a ring of hard bark, the breadth of which varies from location to location depending on how far up the local species of tortoise can reach. There is no reason to expend more energy on this kind of thing than is absolutely necessary. The tortoises, however, develop longer and longer necks to reach up beyond the bark. Some tortoises have developed an extreme elevation of the shell over the neck which enables the animal to stretch its neck even further. It is precisely this formation which causes the shell to resemble a saddle: a *galapagos*, as it is called in Spanish, hence the islands' name. Cactus and tortoise develop in a dynamic equilibrium. When the tortoise is able to reach a little higher, the cactus has to develop a greater reinforcement of its bark. In this manner there is an increase in space for both species without a cessation of development. Deducing what has happened in the Seychelles from developments in the Galapagos is perhaps not a submission of evidence which would stand up in court, but the indication is fairly convincing.

Repetitions in nature are highly diverse. Behaviour and adaptation are a display of parallel situations, regardless of which direction one looks in. The combination of homologous traits or inherited dispositions and environmental influence is indistinguishable from the classic discussion of the causes of human development and behaviour. Are we a product of our environment, or are we locked, unchangeable genetic machines, or possibly a combination of the two?

In nature we can never escape our inheritance, or rather the genes, even if we cannot predict where they will take us. Fundamentally all organisms have the same origin. Some organisms develop because individual cells have swallowed up others, others because the cells work together on the molecular level and still others are a mish-mash of everything, but this does not change the fundamental fact: 'There was once *one* cell …!'

When one examines the genome in individual organisms and compares them with each other one can be utterly terrified by how identical they are. We have much in common with a chimpanzee. That's an idea we've become used to and it no longer strikes us as so incomprehensible, but what about a banana fly? A tiny, primitive, little insect? The human being has around 20,000 genes 2,758 of which we share with a banana fly! We are aware – we think we know – that different organisms' identical genetic structure probably presents us with a picture of the story of evolution, but, beyond this, they don't say much about related characteristics. A single gene can mean a world of difference: just think of the gene for the male sex hormone, testosterone, which is to be found in men and women. A regulating gene ensures that about half of all embryos have a somewhat higher concentration of testosterone than the others – and become males.

If a vacant niche demands the power of flight, flight is invented or rediscovered through the use of the existing genetic material. The available spare parts are taken down from the shelf and the environmental effect forces them in the specified direction, as long as it pays off; as long as each step in itself promotes survival, or merely fails to do opposite. But it takes many steps, species and detours before nature comes up with something as complex as flying.

The situation is a little different with the formation of images in the brain. All the senses – that means the human being's five senses as well as nature's other senses – can, independently of each other, lead to more or less refined

Flightless, the White-throated rail from Aldabra has adapted to island life thus losing its ability to fly.

Land crab (*Gecarcinidae*). Nocturnal scavengers which must return to the sea to sprawn.

images in the brain. It's not the sense organs which do the work, it's the brain, and therefore it becomes difficult to conceive of the brain as an organ just like all the rest. As I said earlier, one can certainly compare survival strategies, which demand the use of the brain, with strategies that function more neutrally in relation to learning and decision-making processes. The success rate does not display great fluctuation, but the question is whether the brain lives its own life and in the long term will change the composition of the species or their forms of expression. The big brain appears to be a flexible organ, which does not necessarily need to wait hundreds or thousands of generations for evolution's answer to present challenges. The brain can do it itself. Could it be that evolution has spawned a monster which cancels out the dualism; the dialectic between inheritance and environment? Has inheritance finally taken over?

It can hardly be entirely fortuitous that we human beings are so preoccupied with dualism: Heaven and Hell, Ying and Yang, spirit and matter, light and darkness, thesis and antithesis, God and the Devil. We are accustomed to dividing everything up into spirit and matter, into inheritance and environment, but, strictly speaking, we do not actually know where it comes from, or even if it relates in this way. Usually we are agreed that either there exist both spirit and matter or just matter. If matter stands alone, spirit springs from it and cannot thus be superior to matter. Absolutely logical.

The philosopher Nietzsche is almost the only one to reject this entirely. First and foremost he thought that it was curious that the human being some thousand years ago began to believe in something non-existent. Neither did he have much time for the alternative, where freethinking results in another faith – the belief in science – which in the final analysis represents an equally inflexible mode of thought.

Currently the foremost representative of faith in science is the English biologist Richard Dawkins, who in his recent book *The God Delusion* attempts by way of logical arguments to repudiate dualism, i.e. the thought that something exists before matter, but the argumentation concludes, (hardly as elegantly as with Descartes) with "God does not exist because he is non-existent". Right, just about anybody can come up with that!

Dawkins and Nietzsche are, however, perfectly in agreement on this last point, but Nietzsche attempts to go one step further towards absolute freedom, which is elevated above faith and nationalist sentiment. One must find oneself, believe in oneself and drop the dialectical nonsense.

A thought is simply valid when it is suited to contributing to the development of life, when it has an effect on the forces in the human personality. It's a matter of focusing on what has value for life, if I may permit myself to modernise the old German a bit.

Nietzsche believed that struggling for truth is a fundamental instinct, which desires to advance life. Instinct takes no interest in the truth as such, but in to what degree the contention is life-promoting.

Indeed, if nothing else, this is a delicate, perhaps even delicious thought with which to play around.

Don't forget the trees!

The fresh resin from the fire explodes in small cracks and breaks the silence, while a swarm of small fireflies rises into the air. Slowly it grows cooler, and everybody moves closer to the blazing bonfire, while tax-free whisky and Old Bitters go round. On the horizon the sun goes down behind thorny acacias which disappear into the dark of night.

The most characteristic icon of the savannah is neither the lion, elephant nor giraffe, but rather the flat-topped, or umbrella acacia which symbolically spreads its crown of thorns out from the trunk to receive the light from heaven. Often it stands in lonely silhouette on the dry grassy steppe like a last bastion against soil erosion, awaiting the next season's lush grass. Meanwhile the crown of the beautiful acacia is clipped neatly right back as if a gardener has been past with the electric hedge-clippers, which however bears witness that a group of male giraffes has passed by and sheered off the fresh shoots. The throat has been stretched and the neck bent all the way back, far beyond the limits of a human spinal cord. The giraffes have eaten the acacia right down.

We lean back to listen and identify the unfamiliar sounds. Everyone is pleased that he or she is able to recognise the characteristic laugh of the spotted hyena, which, for reasons unexplained, sounds like a warning growl of a lion, when later at night one steps out to answer a call of nature. The acacias burn lustily in the big bonfire which, after a while, is so hot that even still-green branches go up in flames. It has been a good day and everybody finds themselves in a euphoric state, in 'another reality' as one of the new safari participants expresses it. The first few days of a safari feel simply too overwhelming and incomprehensible for many people, whose entire experience of nature is often restricted to pretty television programmes – often shot exactly where we are sitting now.

The atmosphere is good. Nobody appears to regret the choice of a couple of good old-fashioned nights camping in the bush so as to be up close to the wandering wildlife: a couple of nights without showers when we don't need to drive back to the hotel to sleep. Everybody admires the orgy of stars in the southern skies, which slowly emerge from the darkening heavens. The romantic quality is muted somewhat by the consciousness of the harsh reality out in the darkness behind trees and bushes. Like another natural law, we experienced hands contribute to the atmosphere with safari tales in which the human being consistently draws the short straw. The steaming teapot goes round, while the faces are lit up by the glow of the fire.

The fire is a necessity but we had a tricky time trying to find fuel today and finally we had to send the vehicle out in quest for some. Much has happened since my first safari when it was possible to gather all the necessary firewood within a ten-metre radius. About midday we saw a truck coming with firewood for Governor's Camp, a big safari lodge not far away. The large lodges have simply emptied the vicinity of everything combustible and now have to scrounge off their neighbours and the areas without lodges. Even though the umbrellas grow fast, it is clearly apparent they are becoming fewer and fewer.

The umbrella acacia belongs to the mimosa family, which is closely related to the pea family. Even though there are many variants, the leaves are often mimosa-like and the seeds are held together in capsules which resemble pea pods. The long evolutionary story has brought into existence around 1,200 species, 129 of which are to be found on the African continent. Acacias grow fast and flower each year, releasing vast numbers of seeds, which are dispersed via the digestive tracts of various animals. This is the precondition for them to be able to spread to new areas and adapt to new niches, and thus the reason that the family is so extensive and dominant compared to, say, the baobab and the many different, slow-growing 'hardwood' trees, such as ebony and teak.

In the Masai Mara we have thirty-eight species, which is a relatively large number in relation to the area's size. In the whole of Kruger National Park in South Africa there are, by comparison, only twenty-one species. Our thirty-eight are spread among different niches from the umbrella acacia in the open regions on the top of the plateau where it actually competes with the slow-growing desert date trees, to the yellow fever tree which is found close to the grey-green, greasy Mara River – in fact exactly where we are based. The yellow fever tree – also called the yellow-barked acacia or just the fever tree pure and simple – demands a lot of water and is to be found, typically around permanent water holes, swamp and watercourses. The first European explorers and settlers became familiar with it as the fever tree because people often contracted the deadly mysterious swamp fever, or blackwater fever, when staying in areas with many of these distinctive yellow trees. Later the fever was given the name malaria (*mal air* – bad air) and even later it was understood that it was some of the mosquitoes which hatched in the water around these trees that brought the fever with them. A good campfire story.

Our African cook and odd-job man has just finished preparing the late camping meal. Everybody throws themselves over the mashed potato, which is made from real potatoes with the attendant Swiss steak and delicious fried vegetables. Dessert consists of fresh fruit and bread fried in syrup. Everybody eats a little too much and awards the cook five stars in the Michelin camping guide. Thoughts go to the humorous Danish novelist Gustav Wied and his creation *The Gluttons' Club*. Some believe that the safari will transform us into very attractive prey.

Shortly after nightfall comes a violent crackling in the bushes around us. Branches snap, leaves rustle but everybody stays calm. About thirty metres from us a large, dark silhouette looms up, tusks first. It's a young female elephant. A fair bit too young to be a matriarch, and is the kind with which one needs to be especially careful. They do not have as much experience and are therefore more nervous than the older animals. She positions herself directly opposite us and stands quite still, while her small flock of seven animals slowly comes forward, spreading out on each of her flanks. The silence is striking, nobody moves, but the distance to the cars is being calculated. For the first time there is nobody thinking about taking pictures, the mouth goes dry, even though we haven't stinted on things to drink. Time stands still while the elephants drink in the darkness. Suddenly it's all over; the herd is swallowed up again in the bushes and we hear once more the snapping branches on the slopes down to the Mara River. We hear them for a long time, the crackling and tumult and some even think that they can hear the elephants munching.

Gradually everyone is getting fairly tired but no one wants to miss a thing, and the experiences from previous nights create no great expectations of a calm, undisturbed sleep. The many sounds, the uncomfortable sleeping-bag underlay, alien surroundings and a good dose of adrenalin are more than unsettling. We knock back three, four shots and exchange more terrifying anecdotes before we finally turn in in our wonderful roomy tent. Yes, just like the good Baroness Blixen, we too certainly have a sense of luxury on the savannah.

It is six o'clock. A wet membrane of dew puts a clammy damper on the morning mood which has just suffered its first check in the recognition that,

Above: At least 73 species of bats occur in southern Africa. This is a large Fruit bat (*Megachiroptera*).
Below: An insect-eating bat of the suborder *microchiroptera*.

Above: Giraffe
Below: Cheetah only hunt on the open plains during the day.
Right: Waterbuck are seldom found more than 200 kilometres away from water.

A leopard makes its way down from the relative height of an Umbrella thorn.

Due to a lack of competition chameleons and lemurs have developed into many different species in Madagascar.

Far left top: Gigantic tortoises of over 200 kilograms once roamed the world but are today found only on remote islands.

Far left below: King protea, the national flower of South Africa.

Left: Elephants give baobabs a hard time but the trees nearly almost always survive the thrashing.

Below: Eating ants.

Baobabs have developed into different species in Madagascar.

contrary to all expectations it can be bitterly cold near the equator. Aluminium thermos flasks of tea and coffee are standing on the embers, and we slink around unwashed and knock back a cup to wake up. Gradually we all assemble on the riverbank in the first rays of the sun to observe our 'private' hippos, which are already in the water. On the opposite bank a large female is on her way down the steep slope closely followed by a half-grown calf. Both raise their heads once and hasten down into the protective river, where they are met with massive grunts from the whole family. In the course of the night they have grazed in the immediate vicinity of the camp, to which the widely spread droppings bear witness.

Traces of the previous night's elephants are conspicuous. During this time elephants really go a bundle on the trees, as they always do in the dry periods. The acacias are toppled, the bark is stripped off and even the medium-sized boughs with a diameter of ten centimetres are swallowed. We see the evidence clearly. Everywhere there are flayed trees, bearing witness to the elephants' ravage, only a few metres from our tents, but this is, after all, the way of nature, and the positive aspect is obvious since the savannah is opened to grasses and shrubs, which are by far the most important kinds of food here.

Unfortunately there is often more focus on elephants' destruction of the vegetation than that inflicted by human beings. Even though one can easily picture situations in which elephants overexploit vulnerable areas, it is the human beings' general 'consumption' which leaves the lasting marks.

Today the trees of the Masai Mara are disappearing at a rate of knots, and it's not necessary to have a scientific investigation to be able to state that fact – it's enough simply to open one's eyes. In the regions at the greatest altitudes there are great, open expanses where the distance between the trees is certainly well over a kilometre. The areas of bush and shrub are diminishing to tiny islands in the open landscape, and the gallery forests along the water course are becoming thinner and thinner without any actual tall trees. In this sense the Masai Mara is much different today from how it was 25 years ago, when bushes, shrubs and tall trees constituted a characteristic icon. The rare hardwood types of trees which have been growing for many hundreds of years, have largely disappeared, and only in the Masai triangle does one see a substantial stand of the beautiful desert date.

The local population is growing, tourism is growing, and where there are people wood is used. Houses have to be built, food cooked and the bathtub has to be heated. Forests disappear. Trees, flowers, grasses, mice, small birds and insects are seldom in the good graces of many when nature is to be protected. It is undeniably far easier to be aware of rhinos, elephants and whales, but without vegetation there is no food for these key species either.

Professional people attempt to take into account the lack of focus on the less conspicuous species by talking about the preservation of nature's diversity, which we also call biodiversity. Nature has, in her own sober-minded way, optimised herself. Seen over a period of time, a natural area nurtures or produces exactly the number of animals and plants which correspond to the conditions at that location. A farmer would say that nature has maximised its hectare yield. The area gives the greatest yield possible against the background of the given circumstances. Areas of land with many species, which is exactly the case with savannah and rain forest, are in reality very stable and can tolerate many changes.

Generally speaking, they continue to resemble themselves as long as the burdens from outside do not become too great. The various species can replace each other, a species of acacia or termite is succeeded by other acacias and termites. In the polar regions it is different, for there the extermination of a single species can set off a chain reaction and threaten every other species. We are thus well guarded, but that is not the same as being able, without further ado, to remove all the large trees.

Africa's natural areas will only exist as long as they can be financed through tourism, and thus it comes across as slightly paradoxical if tourism, through intense use of wood, contributes to the removal of the basis of its own existence.

And the waters parted

We wander along an invisible line, a couple of kilometres in length from the umbrella acacias on the top of the plateau to the camp and the yellow fever trees by the river. The landscape falls some hundreds of metres. Species follows species along the line where the character of the plant life reflects the quantity of accessible water. Water is the most important limiting factor for the vegetation of the savannah and thereby all other life. Light there is enough of, and nutrients remain in the system in that they are only rarely washed out.

Water is not just a matter of precipitation, but is, to just as high a degree, about whether that precipitation is accessible to the roots of the plants. Water actually evaporates extremely fast, so fast in fact that the nutrients remain in the ground. Every shadow, hollow and fissure retains a little extra water and creates its own niche, which is the main reason for the savannah's enormous number of species. The species are, in any case, very unevenly distributed. We know that there exist thirty-eight species of acacia, but only one or two are commonly seen, while the rest conceal themselves in small niches.

All in all the landscapes are most frequently dominated by relatively few species, which may account for more than 80 per cent of the biomass, while the remaining species are rare and inhabit very small, delimited niches. Sufficiently close examination reveals the same state of affairs with insects. There are up to a hundred different representatives of dung beetle and just as many of termites, but ninety-six to ninety-eight of them are hard to find. The few common species are thus overwhelmingly predominant because they are absolutely the best adapted to their current situation, while by far the most species live a stand-by existence, falling into a slumber in their relatively out-of-the-way niches where they wait for more auspicious times.

A single termite has, for example, become specialised in the death of the umbrella acacia and moves in as soon as the last life has ebbed out of the trunk. The dead acacia releases salt, which the termites collect and mix with the cement which makes up the surface of the dwelling. A single species of grass, *Cynodon*, has specialised in the peculiar salt-containing earth of this termite dwelling and moves in after the termites. Both the termite and the grass have small niches which are dependent on a specific action: the death of a tree.

This species of grass is, however, the food of choice of the impala in dry periods, and possibly also the giraffe's. It has, in fact, been shown to be the case that the giraffes actually bend down to gnaw and destroy young umbrella acacias. This is the same principle as the elephants' destruction of larger trees which encourages the production of grass. But destroyed trees may also mean less shade, dry earth and fewer leguminous fruits. The protein-rich pods of the acacia trees constitute a highly prized and valuable source of nutrition for numerous species which make a point of seeking out areas with mature pods during the most difficult of the dry periods, when the nutritional value of the seeds is at its highest. Many creatures can, at extremely long distances, recognise the rustling sound of dry pods falling to the ground. Many of Africa's hunting tribesmen make use of this knowledge, when they throw dried pods onto the ground from their hunting hides to entice game to approach.

After lots of plant stories, we make it back to camp, which in the meantime has been struck and packed away, and drive to one of the Mara River's favourite watering places for observing migrating gnus. On the way we discuss how the savannah is supposed to look. What should be the relation between grass and trees? Some are surprised that one can even discuss the subject since nature seems to be doing very well as it is. But it's not that straightforward, because the savannah is dynamic – like everything else in nature – and it varies from year to year. Much of the Mara was once pure bush savannah with a mass of tsetse flies which can infect cattle with the deadly sleeping sickness, nagana. The Masai have burned off the bush to create expanses of grass and to be rid of the unpleasant flies. Another part of the Mara was marked by the annual wildlife migrations that have also created areas of grass with occasional old trees that have anchored the layer of earth.

All in all the circumstances have created a very varied landscape, rich in wildlife, which we believe worth preserving for posterity. There is every reason to believe that without fire or elephants the Mara would develop into

bush savannah with dense thickets and a low diversity of mammals. Totally devoid of interest for tourists. The interesting savannah thus demands a form of maintenance, just as moorland should be kept free of pine trees.

The varied savannah landscape seems to consist of large, uniform grass areas and correspondingly large dense areas of thicket, but in between these are depressions with water holes and bogs, rivers with dense fringing forest, rocky areas and ridges. Variation offers a profusion of habitats for numerous species of plants and animals, which live in narrow niches, ready to take over new areas when conditions change from time to time. The more variation in topography and presence of water, the greater variation in plants and animals, and thus greater diversity. Great diversity also means greater stability, and thus greater ability to resist in the event of natural catastrophes, climate changes and all kinds of environmental alteration.

Quite how the stability of an ecosystem is to be understood continues to be an area of discussion among ecologists, but customarily it is conceived of as the system's ability to maintain a consistent biomass. Nothing disappears without being replaced by something else. Let us look at an

Above and facing page top: Baobab trees, like acacias: same environment, same adaptations.

example.

As remarked earlier, it is possible to find a hundred species of dung beetle in a large expanse of savannah. Each species prefers dung from a quite specific grazer, even though they always keep a couple of alternatives in reserve. If the savannah is dominated by open areas with many grazers, such as zebra, gnu and Thomson's gazelle, the total population of dung beetles will consist mainly of species which specifically prefer the manure of these animals. If the savannah becomes overgrown, the grazers will disappear and be succeeded by leaf-eaters (browsers), such as kudu, giraffe and the generalist impala. The composition of the dung beetle species will correspondingly be altered.

In conclusion one can say that many of the savannah's species live a kind of stand-by existence from which they may, at any time whatsoever, take over a larger role, if outside factors expand their niche. Hereby the savannah continues to resemble itself; it's just that the colour sounds different.

World war in the treetops

Most acacias have thorns, some long and needle-pointed, others small and hooked like those of roses and blackberries and all with exactly the same purpose as the South American cactus and the germinating palms on the Seychelles. They are to protect the plant from aggressive herbivores, which they do with varying degrees of success. In Blixen Camp the thorns offer good protection against visitors who try to take unauthorised shortcuts to their accommodation, while elephants, rhinos and giraffes don't seem to be put off by this kind of detail. On the contrary, they wolf

A mangrove on stilts, a special adaption to sandy beaches.

down the thorns as if they were the food of the gods. But it turns out that the plants have more strings to their bow.

The slender acacia, which is called the whistling thorn, and a small cocktail ant have thus developed one of nature's complex mutual interdependence relationships. The thorns of the acacia have a powerful, genetically determined tendency to develop a number of large, round galls, in which the ants take up residence. The ants are a hoard of savages that attack everything which comes near the tree – big, as well as small. Beetles, larvae or mammals wanting to take a nip at the tree get a warm reception. If the ants can't get a grip with their extreme mandibles, the unsuspecting visitor gets sprayed with poison.

The ants thus offer the tree good protection. In return the tree responds, not just with light, friendly dwellings but also with highly nutritious food. The tree has some special glands which secrete a sweet nectar, exclusively with the needs of the ants in mind. In the shelter of the tree the ant colonies have developed into a complex society where the individual galls have special functions such as storage space, animal husbandry, breeding space and burial chambers. When the wind travels through the entrance holes in the galls, the tree whistles, hence its name.

The whistling thorn can increase or decrease production of galls and nectar, but it is not the ants that induce or regulate the tree's production. It is, conversely, the presence of large, leaf-eating wildlife such as the giraffe, kudu, impala and pointy-nosed rhino, which cause the tree to react. When the leaves are eaten, the tree's production of nectar and galls is activated. The ants can't reckon on any kind of love-thy-neighbour ethos or altruism from the tree, unless the tree has acute need of their help, i.e. that the production of nectar is not adjusted to suit the needs of the ants. Throughout, the tree functions entirely rationally and conserves as much energy as possible. The ants, however, are so fixed in their behaviour patterns that they are compelled to remain in the trees, and in desperate times with no large herbivores in the vicinity they really have to fight for survival – see below and find out how.

The whistling thorn dominates large areas with apparently dry, well-drained soil where the trees stand a few metres apart from each other, surrounded only by grass. Closer inspection, however, reveals they are often rooted in ground called black cotton earth, which becomes extremely wet and sticky in the rainy season. The ground is, in other words, ill-suited for ant nests which risk collapse. For this reason, many species move up into the trees, where we now know that three other species fight for space against the original cocktail ant, which, for simplicity's sake, I shall call the mimosa ant, after its Latin name. The mimosa ant dominates around 50 per cent of all whistling thorn, but the other species are always poised to try for a foothold, and this has unleashed a regular ant world war.

How can one tell which of the four species came first? Who has the original ownership of the tree, and who are the Nazi intruders in this well-ordered, peaceful society? The cooperation between the tree and the original mimosa ant is of the most exemplary, or symbiotic character, where both species benefit from the collaboration without significant overheads. The adaptation clearly dates from farther back. One of the other invading species utterly disdains the plant's offer of delicious nectar, and instead gnaws off and thus destroys the glands, rendering the plant less attractive to the nectar-eating species. Another gnaws its way across the leaves and branches, removes the growth cells and hinders the tree in building bridges to other trees, from which potential rivals may come. But the tree suffers considerable damage. The last species is not, itself, capable of making holes in the galls and thus keeps parasitic beetles as domesticated animals, so that these can do the job for them. The symbiotic relationship between the tree and the three invading species is thus somewhat strained, or less developed, because they are all wearing the tree down.

In theory it is the oldest and most original cooperation between the ants and the tree which provides the greatest mutual benefit. A long development tends to smooth the rough edges and refine or optimise the relationship. In the final analysis both parties draw most benefit in the same way as viruses, bacteria and parasites do, from the fact that their hosts survive for a long time.

Despite the biting ants, the whistling thorn is the favourite food of the female giraffe. It is clearly apparent that female giraffes seek out areas with whistling thorns, and some believe that they avoid trees with ants, and they can smell the ants' airborne hormones (pheromones) but this is not the case. They eat with relish trees with ants, but whether they avoid the ants, and how they are supposed to do this, remains a mystery. In any event I have never seen them wolfing down ants, but maybe my binoculars are not of sufficiently high quality.

Research has shown that the trees produce fewer galls and nectar glands when their areas of growth are shielded from the big leaf-eaters. This causes the mimosa ant to supplement its food with nectar-eating insects, which they keep as livestock, in the same way that other ants keep green-fly. Like the green-fly these insects are parasites and will damage the tree in the long run. If the tree suffers sufficient damage the mimosa ants will move out of the enfeebled tree, which is relinquished to the other species. Both ants and tree thrive poorly when they are cut off from the natural inhabitants of the savannah, namely the large plant-eaters. Can we, in other words draw the surprising conclusion that the whistling thorn thrives best when it is eaten to a modest extent which compares well with the way lawn grass grows best when it is cut?

This is, to a high degree, a story of the significance of diversity and the complexity of the composition of the species – just one of a million stories about which we haven't the faintest idea. Couldn't we permit ourselves to ask the question, "How can one take a position on something about which one hasn't the faintest idea?" Some years ago, an – otherwise – clever man wrote in a Danish newspaper that he couldn't get upset about the disappearance from

Jutland water courses of a rare fish, the houting. It could be that this is nothing special, but the point is that we do not know for sure. For that reason such a point of view is not particularly intelligent.

The destruction of Africa's trees is a serious threat to the landscape, the ecosystem and the human being's ability to thrive. The most important cause of this is, however, neither an imbalance in nature, nor droughts, fires or any other kind of natural occurrence: it is the human burning of trees.

Without trees, no savannah, no wild animals.

The tree has the floor!

My maternal grandmother had green fingers. According to her own account she enjoyed a certain fame for this honourable laurel-crowned characteristic in the apartment building in inner Nørrebro, Copenhagen, where the unofficial Danish championship in the growing of indoor plants was an ongoing event. Window frames, dark and light corners of rooms, with and without warmth from radiators, were covered in a miniature rain forest. Cuttings were put in water and exchanged together with the magazines *Billebladet* and *Romanbladet* over a cup of coffee without *richs* or *cikorie* (chicory) as the ersatz coffee was called after the beautiful, blue roadside flower which provided its roots for the grounds.

This is a flower about which nobody knows anything, but it is otherwise with exotic rubber plants, mother-in-law's tongue and poinsettias which were tended according to all the rules of the art, with automatic indoor sprinklers, Substral and wiping with damp sponges, while Granny confided to the unspeaking vegetables the most important events of the day, principally the welfare of members of the family. Since my grandmother was fortunately down-to-earth in money matters, a tremendous host and cook and otherwise appeared absolutely normal, one tended not to dwell overly on this peculiarity, even if it was a mite more controversial than conversations with the family's budgie farm. The budgies were well aware of who was in charge, as anybody could see when the colourful birds actually nipped the bread rolls from Granny's lips. Despite my grandmother's Catholic and otherwise rational disposition, nothing could convince her that there was no bond between people, animals and plants. House plants, which did badly, shed leaves or had withered shoots, were stressed. Therefore they should be calmed down by talking to them gently. Judging by the fecundity from the window frames and flower baskets this conversational therapy worked pretty well. But let us look at the African natural world.

Towards the close of the twentieth century a series of deaths on an African farm led to a new understanding of the secret world of plants. We had brought about the unveiling of the complex defence and communications systems, including poison attacks on enemies and alliances with the enemies' enemies.

Jack fruit which has become a staple food where it grows.

The story has its origin on a game farm in one of Africa's driest countries, Namibia, where the hardy oryx antelope, with the long twisted horns, was experimentally bred. The concept of breeding was exceedingly simple: it consisted solely in keeping the antelopes in an enclosed area, corresponding totally to their natural habitat – a dry, sandy savannah region with sporadic acacia vegetation, and in every other respect totally unsuitable for cultivation. The wildlife rapidly began to sicken and grew weaker and weaker until the animals finally started to die as if on a conveyor belt. At the subsequent autopsy it could be stated that the wildlife died of tannin poisoning.

Tannin is a bitter substance which is found naturally in many plants, where it is involved with the formation of proteins; it is non-toxic in small, regular doses, but fatal in large doses. Further analysis of the oryx's primary food source, the farm's sporadic acacias, revealed an extremely high level of tannin, far higher than the naturally occurring level in the trees outside the farm. Thereafter it was relatively simple to verify the findings and conclude that acacia trees are capable of regulating their tannin level in relation to the grazing pressure. The wildlife are permitted to snap up a certain number of the tree's leaves before the tannin production is induced, and the tree becomes toxic.

Tannin causes the leaves to taste unpleasantly bitter, which, under normal conditions, is sufficient to keep the leaf-eaters away, but on the farm the wildlife did not have the opportunity to slip away from the uneatable trees and had to choose between the devil and the deep blue sea: death by starvation or poison.

The strategy of the trees is controlled by the dynamic between the energy-consuming production of toxins and a reduced capacity for photosynthesis when the leaves and chlorophyll are consumed.

The story proved to be more complex. When the trees outside the farm's fences were examined it was found that the nearest ones contained more tannin than the ones at a distance. The closer to the farm, the higher the tannin content. The explanation is that the trees 'under attack' secrete a hormone which is carried by the wind to the nearest adjacent group of trees and there initiates the production of tannin, thus forcing the leaf-eaters to move farther away. The acacias under attack thus communicate the situation onward to the next trees, which then can begin the build-up of defence bastions in good time and protect them themselves from being over-exploited.

This naturally provides occasion for the obligatory question which I raised in connection with the reciprocal altruism of bats: why does the tree use energy to produce more hormone than it needs for its own use? There is no reason, is there, to secrete it through the leaves, for it to be dispersed?

The nearest adjacent trees are probably children, parents, nephews or in some other way related, so that one could certainly venture the point of view that the trees were actually protecting their own gene-pool, even though this might strike some as a tad romantic. There is, however, also a second implication in that the roots of the trees are a final bastion in the fight against soil erosion, about which I have already spoken. In Africa's dry savannah regions the grass is periodically gnawed down to the roots and only the roots of the largest trees can hold the flighty earth, until the rain brings about the next period of seed germination. Once erosion takes a serious grip, all the earth and nourishment disappears. The trees protect, so to speak, their own habitat and safeguard their continued existence.

In recent years numerous examples have been found of complex connections in the world of plants. Research into tobacco plants in Virginia demonstrated that the plants vaccinate each other in almost the same way as the acacias. Both fungal and insect attacks cause the tobacco plants to signal to unharmed plants which almost instantaneously set in motion the production of toxins (pesticides) within the plants. What is even more remarkable is that the tobacco plants also understand signals from other species. If so-called weeds around a tobacco field are attacked by insects or fungi, this is registered by the tobacco plants. Cutting mugwort with scissors leads to the plant sending out an airborne chemical, (methyl) jasmonate (MeJA), which smells of jasmine, and which is also used in perfume. The tobacco plants react immediately by producing substances which give them an unpleasant or 'inappropriate' taste, exactly as the acacias do. The quantity of the substance in question is multiplied by four in a few minutes so that its effect is extremely fast. Scientific research has shown that the tobacco plants which stood nearest to the mugwort had a 60 per cent lower incidence of attack by grasshoppers and caterpillars. It has been shown that many plants produce jasmonate, and that the substance can therefore be added to the growing number of defence and signal substances with which we are gradually developing an acquaintance.

A frequently occurring substance in the plant world is called wintergreen oil. It is found in black currants, pansy, coffee, birch and red clover. Some uses of the substance are in disinfectant mouthwash and as a flavour in chewing gum. It is a chemical compound of salicylic acid, the active ingredient in numerous pain-relieving medicines, such as albyl. When a plant is attacked by fungus or bacteria the production of wintergreen oil is increased, and its presence in the cells is a signal to many different genes to put into production a number of defensive products which can combat the injurious agents.

Recently it was discovered that trees sometimes exude relatively large quantities of wintergreen oil, in the same way that acacias transmit their signal substances, and the purpose is believed to be the same. They communicate with others. The quantity of the transmitted signal substances varies in relation to frost and drought and is therefore related to the current stress-level of the trees. Thus the trees take headache pills, both to reinforce their own immune system and to warn their neighbours.

The communication and defence systems of plants are, however, far more complicated than the processes described appear to suggest. Some plants produce both jasmonate and wintergreen oil, but not simultaneously because the production is controlled, or regulated, by the same protein (gene) which, in principle, can only choose one solution at a time. The one solution, so to speak excludes the other. In professional circles this regulator is called MPK4.

The type of regulator proteins are called kinases and the reason that they are mentioned here is that they are found everywhere in living organisms, from bacteria and yeast cells to human beings. Where there are living processes, these kinases play a crucial role. They ensure that independent cells can communicate with each other. They translate, interpret and transmit signals. The same genes and the same proteins regulate the same processes everywhere in the realm of life – so to speak – and remind us that everything comes from the original stem cell. Or was it a quark?

The plants mobilise

These stories have really forced research to focus attention and the deeper we dig into the plant kingdom the more lurid tales come to light. Let's get stuck into the one about the lima bean, which has expanded its defensive arsenal against a destructive spinning mite. The lima bean sends out aromas or signal substances, which irritate the spinning mite with their unpleasant taste while simultaneously alerting the bean's neighbours. The spinning mite may possibly have learned to live with the lima bean's unpleasant taste, but the spinning mite's worst enemy, the gamasid mite, has learned the lima bean's signal and rolls up to gobble down the unsuspecting spinning mites. Translated into the language of human beings, the lima bean's signal reads: "Leave me alone, look out, come and help, there's some great food here." At least six genes are activated in this response, which is presumed to be initiated by spittle from the attacking insects. Take that!

Thus it is that the plants do not only supplement their outer protective measures such as thorns, barbs and spines, hard, coarse bark with a veritable arsenal of chemical weapons, but they also communicate in lively style with each other and other species in the struggle to survive. The rational explanation is that it is just one of the ways whereby evolution regulates the relationship between various organisms, between primary producers and consumers – the dynamic balance which permits attackers and defenders to co-exist. If everyone is supposed to be here, nobody can be allowed to overdrive. Nobody must take too much. Sounds almost like socialism, doesn't it?

Some plants are permanently poisonous, because the grazing pressure is roughly constant. This can be seen, for instance in the Galapagos, where cyanide is particularly widespread. Other plants only have to produce toxins in relation to the current requirement, as we have seen with tannin, so the plant or tree does not expend energy on developing unnecessary poison. The tannin poison is moreover 'the recycling' of a substance which is already in use for other purposes, so things could hardly be more rational.

Strong chilli is a corresponding example. Various species of the chilli fruit develop varying strengths for defence against importunate insects, but some species grow appreciably stronger when they are attacked. Therefore the 'insect-attacked' chilli fruits are a particular speciality among enthusiasts who cannot get things strong enough.

Before collapsing in a total faint at how the many fantastic contexts and adaptations in the plant world have developed, one should bear in mind that their short generation span in comparison with that of many mammals, combined with the vast numbers of seeds (progeny) in itself, contains the germ of a fast and, in many instances, complex development..

If a seed is to survive a savannah fire or the digestive juices of an animal, because they are often, respectively, burned or eaten, there may well be an instance of extremely rapid adaptation. If most of the seeds in one generation are eaten or burned in a savannah fire, only the hardiest seeds with thick capsules will survive. In the succeeding generation there will be a significant preponderance of seeds exhibiting this characteristic. Here it is precisely a matter of the evolutionary mechanism of rejection, which I have already

Verbena bonariensis enjoyed by two femail Monarch butterflies.

Swallowtail butterfly on an invasive bouganvillea.

explored. The worst-adapted are effectively weeded out, and force the organism in a completely determined direction, as long as the new characteristic can justify itself.

The diametrically opposite situation is, however, just as likely. If the large seed-eaters are eradicated and the savannah fires are fought back, the thick-husked seed capsules will have a problem with germination. Many seeds will rot, but the surviving thin-husked seeds have gained a significant advantage because they are now permitted to germinate without being destroyed by digestive juices or fire. There is no doubt that the plants are capable of reacting swiftly – to the joy of the organisms which depend on them.

Rapid adaptation is presumably the plant world's most important survival strategy. Toxins, adaptation to savannah fires and digestive juices are, seen in isolation, simple evolutionary adaptations, which can be developed fast. If these are combined with communicative enzymes which signal to neighbours and other species, and set many different processes in motion simultaneously we are in a completely different ball game, and can see the result of a long-term evolutionary development. The interesting aspect is, however, that these very complicated processes are often regulated by a single protein which blocks, suppresses and sets in motion all at the same time. The unravelling of these processes ends up taking a very long time and, for now, we must be content with stating that evolution appears to recycle the available proteins in various new situations and increases or diminishes their status, all according to requirements.

The benign poison

Kigelia is the local name for another of Africa's very characteristic trees with large, dark, red flowers, which, in the same way as the baobab, is pollinated by bats. Each tree stands in near isolation, spread across the Masai Mara and Serengeti and the whole way south through the African continent to South Africa, where sporadic night frost sets a limit to its distribution. The long, cylindrical fruit can weigh up to four kilos and hang from long stems which cause the tree to resemble a string of sausages in an old-fashioned butcher's shop. Therefore they are known by most Europeans as sausage trees.

With fruits on it, the tree is quite photogenic, but we tend, as a rule, to approach it with caution, for there may be a leopard concealed in this tree, standing apart and alone. David Livingstone was one of the first to describe this curious tree, under one of which he camped on his way to the Victoria Falls at the point where the modern countries of Zambia, Zimbabwe, Botswana and Namibia meet. Another of the sausage tree's local names is *kazungula*, which translates as 'the sheep's fat tail'. Favourite children have many names.

The tree's thick bark and extraordinarily tough wood offers good protection against insect attack. The fruits and, in particular, the seeds are moderately poisonous and are often left to rot on the ground. Even so, both giraffes and monkeys nip a couple of mouthfuls now and again. And the grass-eating hippopotami apparently like to route their nocturnal wandering by way of a sausage tree to gnaw a bit of the fruit. It is possible that the tree's toxins play a role in connection with the animals' digestion, or, also, they consume it as medicine to rid themselves of unwanted parasites.

In nature there are always species which prevail, break down or exploit the poison of other organisms. Both the black rhino and the kudu antelope have, for example, developed an enzyme which can break down the massively toxic white juice from the candelabra tree, a few drops of which would be lethal for human being. The human being, however, learned a long time ago how to detoxify the seeds of the sausage tree by roasting, and most African tribes have kigelia seeds on the menu when other, more easily accessible, food sources dry up.

We stop in front of one of the Masai Mara's few sausage trees. As usual there is no leopard but the humid area around the tree can offer shining water and a rich bird life. Spoonbills, hammerhead birds, and saddle-billed storks have found their way here. Hundreds of metres from here are half a dozen vultures of various species in the middle of polishing off the last remains of an impala, while a marabou stork and a couple of tawny eagles are sent to the back of the queue. There's a lot to see, even though it is always exciting to catch sight of a few more strange spots in the tree that could be a leopard. In the meantime I tell the story of the sausage tree and poisons. A bit odd, doesn't it strike you? All these poisons in nature? But nature isn't lying in wait for people. Poison is a human explanation for something we cannot tolerate and which has never been 'intended' for us.

Some snakes, spiders and scorpions use poison to kill their prey, but the poison is produced primarily to protect and warn, seldom for killing. It is in fact unnecessary to waste energy on killing something that won't be eaten. The latter is primarily a human enterprise, as, for example, Augustus's beloved wife Livia was well aware when she killed her husband with the juice of deadly nightshade which was smeared on his figs.

Both people and animals benefit from nature's poisons and have learned, moreover, how to defeat toxicity. We only need to look at the fields on the periphery of the national park where one finds a plant with long, thin stems and a fan-shaped crown of leaves. In fact we need to look under the plant for its most important component, namely the large, poisonous, swollen root which is known as manioc or cassava. Cassava comes from South America, where the Indians learned how to clear the jungle and cultivate the plant which is today the most important staple food among the self-sufficient farmers of Africa and South America. The swollen root is fairly poisonous and has to undergo a lengthy process in which it is washed, left to steep, dried, grated and roasted. The end product is excellent, highly nutritious flour, filled with carbohydrates, which is used for the making of bread and porridge. How our forefathers in the Amazon worked their way towards this highly complicated process around 5,000 years ago, is still, and will be for a long time yet, something of a mystery.

In Denmark and many other European countries the beautiful foxglove is a brilliant example of the human being's exploitation of plant poisons. Known in English as 'foxglove', the plant's Latin name, *digitalis*, refers to the two-year herbaceous perennial's bell-shaped, thimble-shaped flowers. It is found in many places with nutrient-rich soil and has been cultivated as a medicinal plant since the Middle Ages. The plant is extremely poisonous and contains two so-called glucosides, digoxin and digitoxin, which can kill both horses and humans, as easy as anything. In small quantities the active substances can make the heart beat more strongly and, at the same time can slow down the heart-rate. For many hundreds of years the extract of the digitalis flower has been the closest that mankind can come to an effective heart medicine.

Another important medicinal plant is the opium poppy, which contains an excess of partly toxic alkaloids with medicinal terms such as morphine and codeine, under the collective name of opiates. Through the years, opiates have brought humanity many joys and even more sorrows. The ambivalent coexistence with the plant is of many years standing and may in fact date back to the birth of civilisation and the cultivation of plants. The Sumerians called it 'the plant of joy' and the ancient Egyptians used it as the upper classes' medicine of choice. Christianity was, however, relatively swift to condemn the medicinal use of the plant since sickness was God's punishment and should thus not be alleviated – a notion, moreover, which has left its mark on the history of the religion and played a not unsubstantial role in connection with the witch trials of the Renaissance. We are not familiar with the original opium plant, since the seeds have travelled with people in the civilised world for the last six or seven thousand years, during which time it has been distributed, new variants have been bred and improved, distributed again, and so on. The opiates were presumably a part of the plant's original defensive apparatus, but today's sky-high content of opium is a result of the human being's 'improvements', from which the plant itself has hardly benefited.

Why is it that the human being has been so attracted to these special plant poisons through the ages? Why have we bred them, improved them and dragged them around with us regardless of where we've been? Because they are perfect imitations of our own self-made 'reward proteins', e.g. endorphins, which cause us to feel a sense of well-being in a wide variety of situations. Our genes, in fact, contain the code for these substances, which are grouped under the general heading of neurotransmitters, and which are found primarily in the brain, where they presumably control much of our behaviour. Simply put, we undertake the actions which give us the greatest satisfaction and each of us is from the time of our birth a kind of drug addict dependent on well-being.

The opiates resemble and also appear to work in the same manner as neurotransmitters in the human brain. But why are they found in a poppy? What is the link between us and the poppy? We do not know the answers, but

A kudu with a magnificent set of horns.

when we take drugs originating with the poppy we are cheating on the scales, and we cannot foresee the consequences. In any event the *bricoleur* has been at work again. Recycle what can be recycled. A small quick evolutionary pull and the function is altered. Poison and patchwork quilts. Anything goes.

Root in the intranet

For most people, nature consists primarily of the visible. Big trees, pretty flowers, large animals such as the elephant, maybe a bird, and, at a pinch, an insect or a lichen. Even though a million and a half stampeding gnus are a ferocious pile of meat or biomass, they do not represent the weight in the ecosystem. That is right under our feet, down in the ground, and is made up of roots, roundworms, amoeba, fungi and bacteria.

The greatest part of the communication between plants takes place underground in the root system, and from there on out to the surrounding environment which is an independent ecosystem, in as much it is meaningful to talk about delimited ecosystems at all. Many plants propagate via rhizomes and remain connected, which, for example, is the case with strawberries that send signals round in the rhizomes or intranet and communicate over great distances. On the one hand the plant can send warning signals out into all the out-of-the-way places and initiate defence processes, but on the other hand, harmful microorganisms can also spread throughout the same system more easily.

In all higher organisms – organisms which consist of more than one cell – there is need of communication between the cells. Plant cells have various functions; some, for example, become fine veins or small channels, which can only be seen under an electron microscope, and through which the plant sends proteins (signal substances) and sugar substances. If a cell is damaged by attack from outside, it shuts down, for instance, a signal substance which is, in itself, a signal. The same signal substance can increase in other situations. The same signal substance can, thus, have multiple possibilities of interpretation, even though one cannot immediately designate this as syntax.

Plants can respond to bacterial attack from the soil by sending out signal substances which attract benign bacteria. These bacteria get houseroom and waste products from the plant and repay this with the formation of a complex protective membrane, a so-called biofilm around the plant's roots. The film is very resistant to attack. Plants under sustained attack will often mobilise a complicated response which involves many genes, which both shut down activities in the area under attack and drop the photosynthesis in the leaves, while the forces are used in other locations.

The roots of a plant do not simply stand in soil. They are in fact located in the middle of a formidable flora and fauna, which almost completely enclose them. In a single gram of soil there can easily be more than ten billion bacteria and a million fungal spores together with a hundred thousand single-celled animals and plants. The activity of the roots – the exchanging of substances with the surroundings – means that the concentration of microorganisms is twice or three times higher in the immediate vicinity of the roots. These enormous numbers of microorganisms are all, in one form or another, related to the plant's activities. The thriving of herbivores is bound up with that of the plants, which is bound up with that of the microorganisms and vice versa.

The gigantic number of microorganisms may not say a great deal in itself, but if one realises that a teaspoonful of soil contains just as many organisms as there are people, mammals and trees on the planet, one would be getting pretty close. It's not a random bacteria soup but a delicate ecosystem with thousands of species, clans, groups and families which, in the same way as the population of the Earth, are in one or another dependent relationship with each other. It can be near or distant, friend or foe, temporary balance or imbalance. There is cooperation, communication and the making of mutual agreements. One lives off each other's waste products or they just eat each other. Once in a while they fight each other until a new dynamic balance emerges which safeguards the existence of some or maybe even all on another level. Everything takes place in the teaspoon and next to it is a new teaspoon, and another, ad infinitum – all linked to each other. It happens in the Masai Mara, Yellowstone as well as in Køge Bay, just south of Copenhagen.

Even though the microorganisms are small, space can be cramped, because the individual populations grow as much as they possibly can. A fungus competes with bacteria which come too close. The response of the fungus is a poison which is found in a small biological barrier around the fungus. The poison kills the bacteria and keeps them at a distance. Finally there is peace. The fungus is able to stay where it is, and the bacteria can operate outside the safety zone of the fungus.

The fungus grows and takes a bigger bite of the cake, the bacteria are killed in their millions, but suddenly a single bacterium survives, because in that one among billions and billions a mutation has emerged. The small, new change in the bacterium's proteins (enzymes) is simply capable of clipping the poison into bits so that the bacterium can now devour the individual inactive 'stumps'. This is rather what happens when a human being boils or roasts poisonous vegetables or when the kudu antelope and rhinoceros break down, in their stomachs, the poison of the candelabra tree. Nature uses enzymes, we use heat. The new resistant bacterium experiences halcyon days, because its fellow species have been wiped out and the fungal poison is transformed into a new food resource. The new clan grows and takes over the entire vacant niche, where it can, in theory, live unnoticed.

The bacteria can communicate with their neighbours because they often need each other. Bacteria, fungi and single-celled animals take care of various stages in the breaking-down processes of nature. Among other things they take care that the nutrients in old pieces of plants can be reused by the living roots and their plants. A bacterium can thus be dependent on the thriving of a neighbour, because its products are exactly what it should live off. To put it briefly, a bacterium can thus be interested in protecting other species. Now if the kind neighbour is attacked by a poisonous fungus, there is help at hand from the resistant bacterium which simply puts out a pipe-line and sends a copy of its resistance gene to the neighbour. One species can transfer its resistance to completely different species. With bacteria the resistance genes are packed in small round rings, plasmids, which can duplicate themselves, independent of the bacteria cells' remaining DNA. This is merely one of the possibly infinitely many examples of meaningful communication in nature.

A wise man by the name of Dr Flemming discovered by chance that fungal poison could kill bacteria in a petri dish culture. He speculated over whether the poison of the fungus might have the same positive effect on the bacteria which made human beings ill. It did, and humanity was enriched by yet another wonder substance from the world of nature. Penicillin was discovered and celebrated great triumphs the world over. The success of penicillin is so great that most of the world's six billion people use it at various times throughout their entire lives, that is, if they can afford it. So do the thirty million pigs we produce every year in Denmark.

When we take penicillin exactly the same thing happens as in the ground. Often all the bacteria are killed, but once in a while a single one survives. And so here we are when the trouble is just beginning. A great deal of the human being's enormous consumption of penicillin ends, for instance, in our waste water and a drop of wastewater here and there kills billions and billions of bacteria except one. Once in a while two bacteria with different resistance genes meet and 'decide' to exchange plasmids, which they often splice together to make a simple plasmid with two resistance genes: it is, after all, only practical in these troubled times to be protected against as much as possible. The meeting between resistant bacteria can take place everywhere, including in the human intestine, where bacteria live with many different resistance genes. The chance that the resistant bacteria encounter each other in the intestine is not so great, because the distance for microorganisms between the beginning and end of the intestine is just as long as the distance between the Earth and the Moon is for us.

One day most of the intestinal bacteria are killed by penicillin, except for the resistant ones, of which some are multiresistant. The distance between these bacteria is for an hour very short, because the enormous barrier of other bacteria has disappeared, and the probability of the exchange of different resistance genes increases observably. Suddenly one is a carrier of a very hardy, multiresistant bacterium, which need not be an immediate problem, but when one is ill with a harmless pneumonia one must hope that one's resistance bacteria remain on the side of the host.

The fear of resistance to penicillin is more than half a century old, and an understanding of the mechanism which brings this development with it is nearly as old. It points, if cautiously, in the direction of a redefinition of the concept of human intelligence.

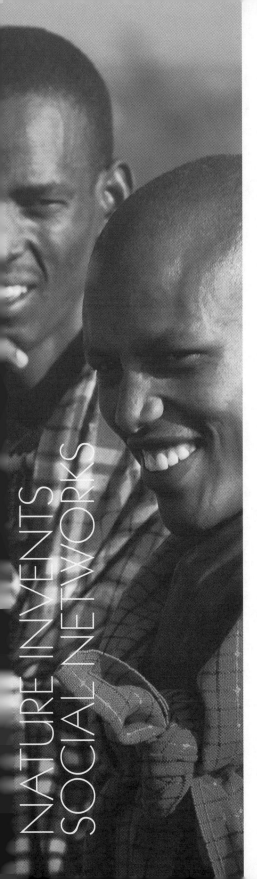

NATURE INVENTS
SOCIAL NETWORKS

Society is a compromise with freedom.
All freedom has its price.

In the Cradle of Humankind

It is one of those days when Thor's chariot rolls across the heavens with apocalyptic force, while he swings Mjølnir hither and thither. I count the seconds and wonder how it is possible to calculate the distance between sound and light when one clap of thunder overlaps another and the lightning stands quivering on the horizon like a glow lamp. The summer rain of the Limpopo is always accompanied by threats of thunder and lightning. Often threat is as far as things get, but not today when the rain crashes down over the grass-thatched half-roof from where I observe the red sandstone monolith of Entabeni on top of the characteristic Hang Rock in South Africa's Waterberg Massif, with mountain on mountain, stretching way out into the forest.

The first rainy day of the season, and everything breathes a sigh of relief. The Earth and the plants suck into themselves like bath sponges, but there is water enough to fill the brooks and water holes. The reservoirs must be filled; it is a long time until the next rainy season. For days farmers, agricultural labourers, game wardens, and pretty much everybody else have kept an eye on the dark, threatening horizon. Waiting impatiently and without a romantic gaze on the fantastic scenery, which presents itself from time to time, when the rays of the afternoon sun are broken by the drops in the black clouds, or when the sparks of the sky light up clearly behind the dry seed capsules of the sugar bush.

Now there is grey in the grey: the teeming rain redeems numerous brooks which fall silver-glinting down over the steep walls of the massif and the word Waterberg – Water Mountain – suddenly acquires meaning. Water and life are communicating vessels. It is life which streams down over the sandstone; the place is sacred. For thousands of years, prehistoric humans, Bushmen and San have made this seeming 'detour' on their route in order to hunt, gather berries and plants or fetch water. Generation after generation has been drawn to the Water Mountains, which have had their name inscribed in the collective memory of the human species through mythology and religion. The early people have diligently paid for the riches on protected mountainsides and in rocky caves with wall paintings that bear witness to the first cultural insights of early man. How they arose, developed and subsequently became ourselves, is something of which we have only a vague idea. For this reason I permit myself to speculate a little further in what follows, where I am granted an intuition which goes beyond the purely scientific.

The South Africans call a little area southeast of here 'The Cradle of Humankind' – because of the discovery there of extraordinarily large quantities of human bones and tools. More here than at any other place in the world. There are two reasons for this. One is that there has been considerable human activity for millions of years, and the other is that the geological conditions in this precise location are capable of preserving the ancient organic material. How much the development actually started here or in East Africa, nobody knows, but there is no doubt our forefathers wandered backwards and forward many times, possibly for a hundred thousand, possibly for millions of years.

When we lay their bones and skulls out in various sequences, according to height, brain size or body height a picture of our development emerges, but it is far from unambiguous. Once we believed that the sequence observed a direct correlation with the size of the brain, but this is far from the case. There have been ancestors with larger brains than their successors, but why this was, and what they used them for, we do not know. Could it be that they mixed with each other after many thousand years of separation and made the picture muddier than we expect?

A few weeks after I wrote the above lines, the natural sciences journal *Science* published a research article which maintained that we – the modern human being – actually have genes in common with Neanderthal Man, who died out 30,000 years ago. Absolutely counter to what was previously believed, DNA analysis shows that we have mated with each other and mixed genes at a late stage of development, even though we were originally separated from the Neanderthals around half a million years ago. Despite this long separation we had not developed so far from each other that we could not mate with each other and have fertile offspring. The human being is thus also a Neanderthal!

At first glance it does look fairly unlikely that we should have lived side by side for many thousands of years without mating with each other.

It's a curious feeling to find oneself here in the cradle of humankind and think that the mountains and the sugar bush stood here many years before the arrival of our ancestors. Do we even mean anything to the plant? It has just opened its seed pod. The kernels lie on the ground and absorb the first rains, the definitive signal to the inside of the plant cells that germination can begin. The tree-like bush with the big red flowers has stood here a long time. It has adapted to the sub-tropical alpine climate of the Waterberg, and it is probable that our ancestors licked its sweet juice, just as their descendants much later extracted sugar from the plant. People have always felt attracted to what is sweet and rich in fat, that which contains energy, which can cause a bit of a problem in a civilised age, where the demand for energy has visibly declined.

The wise Linnaeus called the sugar bush's family Proteaceae, after the Greek sea god Proteus who could foretell the future. Fortunately the god could also change shape and thus escape stupid questions posed about the future by the inquisitive. Linnaeus's assigning of the name refers to the many forms of the sugar bush family, which, in fact, he knew only from a very few, rare drawings brought back to Europe from the Cape by Dutch sailors.

The lovely protea family with its something under a hundred local species constitutes a considerable part of South Africa's special community of plants called fynbos, which is the world's smallest, but curiously enough displays the greatest diversity per unit of area. A square metre of fynbos contains more species than a square metre of rainforest. The plant community's name comes from the expression 'fine bush' which refers to the plant's beautiful wax-covered surfaces which can be found in grasses, flowers and trees. The plant's surface resembles that of holly with a lovely glossy sheen and this wax layer is also to be seen on the species lyme grass among the sand dunes of Europe.

In all the plants the layer of wax serves as protection against evaporation and dehydration. The 'oilcloth strategy' is thus at home in areas where the accessible bodies of water are few and far between and the danger of dehydration ever present. The coastal stretches of South Africa and Western Europe, with their parching wind and highly porous, sandy soil, are obvious terrain for this strategy. It is precisely in this way that the fynbos flora has developed. Fynbos actually consists of plant families that are found all over the world but here in South Africa every local variant subject to the same environmental effects has developed exactly the same survival strategy. Might we not be entitled to assume that this is another case of homologous qualities? The same proto-genes are activated in the various families in the course of adaptation to the same ecological conditions.

Fynbos is the world's sixth and smallest community of plants and covers, in its most prominent form, an area along the coast of South Africa just a little larger than Denmark. From here it puts out shoots through river basins and ravines into the dry highlands where the wealth of species diminishes and special adaptations create new species, one example being the sugar bush.

Adaptation to water is found all over the world and therefore the adaptation to fire is the fynbos flora's most interesting characteristic icon, and also the one richest in perspective. Masses of the plants around the Earth's savannah belt have – as I mentioned earlier – a comprehensive adaptation to fire, but in the case of the fynbos we can see this development in its most extreme form and the frequent occurrence of fire has pushed the evolution of the flora to a massive diversity. Every plant seems to have its own answer to the problem, which is no longer a problem, but a precondition for the flora's existence. Among members of the protea family some submit to burning to the ground, after which they, unaffected, put out shoots from the bottom; others have a thick, fire-resistant bark, and many have the same strategy as the sugar bush: they bring forth robust wood-like flowers that dry and close themselves protectively around the seeds which are only dispersed one year later when it rains.

Roughly one third of all the fynbos plant species receive appropriate assistance from ants. Over the seed the plant deposits a fat-containing substance, a so-called elaiosome, which makes the ants drag the seed down into the dwelling where they feed the larvae with the 'fat' which the seed itself does not need. Thereafter the seed lies safe and snug at the bottom of the ant's nest, protected against fire and waiting for the rain. Germination in the case of some of these seeds is initiated by smoke, and in the case of others by heat. The relationship between the ants and the dispersal of seeds is called myrmecochory and it is, moreover, fairly frequent. It looks as if the form of seed dispersal acts as a catalyst in the plants' evolution and brings about a quick and effective adaptation, possibly because the selection mechanisms are powerful. The ants gather the best seeds. The best seeds germinate, and so on.

In South Africa lots of time and trouble go into caring for one's small, unique plant community whose remaining areas all enjoy one form of protection or another. Every management plan contains a 'burning' programme. The fynbos has to be burned off at very regular intervals if diversity is to be maintained. The burning should preferably be a so-called 'cold fire' which moves rapidly through the scrub and 'lightly fries' the surfaces without leaving serious damage. Why cannot the areas look after themselves? What kind of fire shaped the landscape previously, and which is not to be found today?

The question has been dealt with in numerous dissertations, and I have discussed it for years with nature conservation people who, more or less consistently, put forward somewhat flighty theories about fire naturally resulting from lightning, lens effects and processes of fermentation in nature.

Just supposing we accept the lightning idea – most lightning is accompanied by heavy rain which puts out the fire. In more than twenty-five years I have observed thousands of flashes of lightning and lightning bolts, but never have I seen a naturally occurring fire. I know it happens, but it is definitely not an everyday occurrence. When TV shows dramatic pictures of forest fires in Greece, California and Australia, these are nearly always accompanied by pretty solid testimony of the fires having a human cause, either intentionally or through negligence.

The fire comes from people. Our ancestors probably affected the landscape and the plants for more than a million years. The organisms form each other and develop in parallel, and we have probably shaped our surroundings through the use of fire. We are aware that evolution can progress rapidly, when the pressure is on, but we have had oceans of time to leave our mark on the world, and my guess is that we have not only played a considerable role in the development of the fynbos flora, but also in that of the savannah and its many species.

There, where archaeologists have lifted bones, tools and food remains from our ancestors out of the earth, one can also find the sites of some of the world's oldest fires. They are between one and one and a half million years old and are thus contemporary with our ancestors' development of better, more refined and effective stone tools. Stone tools and fire represent an enormous leap forward in human history and probably form a link to the ancestor we call *Homo erectus* who began to dominate extensive regions at this time, and later migrated out of Africa. Their ancestors within the human race, whom we call the Southern Ape or *Australopithecus*, were forced out by the competition, apparently at the same time.

We can, naturally, never know whether these prehistoric humans could make fire themselves, or, rather, if they discovered how. They learned very early how to pick it up and keep it alive. They experienced naturally occurring fires which they learned how to seek out to find dead animals, killed by the blaze. The smoke on the horizon was a blessing, for it indicated where there was food and fire. They learned that fire gives warmth, it kills prey and roasts them, so that they are easier to digest. How difficult could that be for a good human brain?

The evolution of the human being staked itself on a complex brain, which enabled the humans to respond differently to different situations and challenges. The brain compensated or over-compensated for the missing physique, but even so, many human species went into the crucible of evolution until one species suddenly took over and began to control nature and evolution. In the first instance fire was the tool that pushed the human being and its evolution forward.

Impalas, ever alert, observing a leopard wandering past.

Today most indigenous people living in close contact with nature have well-developed methods of carrying around living fire, which is laboriously packed in mosses and carried in calabashes. Fire is something watched over and shrouded in numerous myths and rites which neither matches nor once-only lighters can change. Fire is holy.

Among the Masai of East Africa, the job of making fire is the responsibility of the men, but the care of it belongs to the women. If the fire goes out then all hell breaks loose and the damage cannot be undone with a matchstick. Instead the shamed woman must humiliate herself by going to beg for some fire from a neighbour. It must be the same fire, the tribe's fire. Fire and fire are two things: a notion which probably has its roots in a time when fire was hard to obtain.

First and foremost, control of fire offered new possibilities for hunting and a whole new protein-rich diet. Using fire it was possible to encircle large areas, in which many animals perished and could be gathered up, and with more refined strategies fire could be used to drive prey into a trap. The same procedure was followed much later by the Prairie Indians when hunting buffalo. The use of burning-off as a strategy has obviously been a factor through hundreds of thousands of years and continued by much later nomads and practitioners of slash-and-burn agriculture. The burning by human beings

has probably shaped large parts of the natural world, from the Danish heath-
land, South Africa's coastal regions and the East African savannah to the
Amazonian rainforest. Thus special high-altitude photographs of the Amazon
reveal patterns in the vegetation which suggest extensive human activity.

Burnt and half-roasted animals inspired our ancestors in the development of
the culinary arts. Roasting/frying eases digestion and came to appeal to the taste
of human beings, even though it's a moot point which came first: the taste or the
technique. The opportunity to replace a predominantly vegetarian diet with a
powerful, protein-rich one has been in itself a massive leap forward but when it's
combined with roasting and boiling then we are dealing with something which
is, in a literal sense a revolutionary development. To put it simply: less energy
is required to digest what one eats. Effective intake of nourishment means that
one can eat less and thus catch less; there is more and more energy to withstand
cold and illness. Fire has naturally also brought with it such benefits as heating,
light and protection from wild animals that, however, would be reckoned
fringe benefits in comparison with the aforementioned. With fire our ancestors
caroused in luxury, and the harsh earth became a pure Land of Cockaigne for
the enterprising ape who slowly began to discover spare time. Where animals
need all their time to eat, digest, mate and care for their young, with the human
being there developed a small pocket of time, a surplus of time.

The human being began to think and develop communication, even though actual speech lay millions of years in the future. They had time to improve and refine their tools, develop types of hunting and hunt in larger and more effective units. Whereas their forefathers had to live isolated from each other to minimise competition for food, they could now live in larger groups which meant better protection for their offspring and ensured better provisioning for the entire community.

The first ethical concepts could have arisen in these circumstances as simple rules which regulated human co-existence. Our early ancestors had to be sure that their offspring would not be eaten by other members of the group, that everybody knew their position during the hunt, that some looked after the fire and others again took up the fight against intruding enemies.

The human being began to organise itself into a society as soon as it was in a position to do so, exactly as independent cells and many species seem to organise themselves on higher levels when it is possible or beneficial for the species.

The story of the chicken and the egg

The cyca, or cycad, stands on stony ground, and the rain-wet, stiff 'palm leaves' project on all sides, completely unbothered by the pouring rain. The sturdy, bowing trunk lies along the ground and is testimony that this specific plant may have stood here for a couple of thousand years. The species itself is, however, between 200 and 300 million years old, and together with the crocodile is one of the relatively few relics from the primeval continent Pangaea, i.e. from when the Earth went out of joint and the continents split apart, first into Gondwanaland and Laurasia and later into the continents we know today. If a living fossil could talk in a way we could understand there would surely be many heroic stories of Southern Apes, *Homo erectuses* and Bushmen on the menu, but for now we must content ourselves with deciphering images and interpreting those tools and bones left behind.

Cycas is one of the first higher plants which developed at the same time that the first herbivores moved inland, a long time before there was any thought of mammals and human beings. At this point, the primitive single-celled proto-bacteria had undergone several billion years' development in the sea. They were now genuine cells equipped with nuclei, walls and sexual reproduction and, finally, some of them became the complex multi-cellular organisms which later crept onto terra firma and developed into cycas and human beings.

As yet, the cycas had no bat or bird to take care of the pollination tasks, but had to develop a technique according to the principle of the available nail, as can be seen today. When the pollen of the male plant is ready, it raises the temperature and emits a powerful scent which drives visiting insects out of the cone with male flowers and onward towards the nearest female plant, which correspondingly emits delicious, attractive aromas. As with modern plants, roses say, this is also a case where the insects drag the pollen around with them, even if the procedure is somewhat complex, and the plant represents an early stage of development.

We imagine that single-celled plant organisms gained a foothold in the tidal zone and gradually accustomed themselves to an increasingly dry environment which demanded that the plant cells developed walls with a layer of wax over the cell membrane, so they could resist dehydration. We are familiar with this layer, the cuticle, which is today rediscovered in the fynbos plant species and many modern species too. The wax layer also required the development of openings, guard cells, so plants could breathe.

Adaptation after adaptation made it possible to take over new territory. Some adaptations also brought about limitations, for instance the free mobility of the cells was lost at the same time that they gained the hard wall. On the other hand the new cells were good at stacking up on top of each other where they began to collaborate. They signalled to each other about executing different, coordinated functions and gradually became plants. The plants resembled the ferns of our day and spread with the aid of spores, which today are adapted to relatively high humidity. It was only later, when seed plants arrived that the completely dry areas could be taken over.

The plants were developed side by side with the animal cells. These retained the soft cellular membrane and had to stay longer in wet or damp surroundings where they developed tentacles, cilia or tails which provided movement and propulsion. Thus the foundation stone was laid for the two cell kingdoms: the plant kingdom and the animal kingdom, each with its own strategy. The plants gained roots and had to come to terms with staying in the same place for their entire lives. On the other hand they have invented ingenious ways of dispersing their offspring. Rosebay willow herb and dandelions have, for instance developed the pappus for their seeds, so the wind can spread their progeny, but species such as the cleavers and avens have developed crooked thorns which stick to and hang on the animals' pelts and, as mentioned, numerous plants spread their seeds via the intestines of birds and mammals, via water courses and the waters of the oceans as well as in the mouths of ants. Animals move more or less freely around and can seek out their own niche, while the plant seed really needs an outsider's help if it is to avoid falling on stony ground.

The first signs of life of which we know are about 3.6-billion-year-old fossil spores of micro-organisms. These first organisms lived on an oxygen-free planet, and much suggests that they drew their energy from reconfiguring various sulphur compounds. Many of their nearest extant relations are so-called thermophile organisms which feed themselves the same way today. They are often found in extremely hot locations, e.g. hot springs, hence their name.

Later in the Earth's history, about 630 million years ago, there developed photosynthetic microorganisms, which probably resembled the extant cyanobacteria.

Like all other photosynthesising organisms they had the capability of exploiting sunlight as power, and oxygen was thereby liberated into the atmosphere with catastrophic results for all life at that time. Oxygen is in fact a violently reactive molecule which in many respects is poisonous and must have threatened all life, until life took other forms and learned to protect itself against it and even put it to use.

Every time the single-celled organisms developed in a new direction, some remained behind in the old niche and developed in another direction. Thereby arose the first large diversity among the species of the planet. Some species developed towards higher levels, such as fungi and protozoa. They began to live off each other, communicate with and signal to each other, which gradually resulted in multiple-celled species. Very ordinary lichens, which we find on the trees and stones of the forest, are a good example of some of the processes which must have occurred in the dawn of time. Lichens are, in reality, two different organisms, which work closely together without being so integrated that they cannot be separated in a laboratory. Lichens are algae surrounded by fungus. The algae contribute to the joint venture with photosynthesis and the fungus keeps the algae moist and provides it with nutrients from the surroundings. Combined they constitute an extremely hardy organism, which can take over nature's most inhospitable niches, where they would never have had the chance to survive independently. It is precisely in this manner that science believes the first multi-cellular organisms emerged.

The symbiosis between algae and fungus recalls functions we know from advanced cells. These types of cells have a cell nucleus, where the genome is stored, and many different cell bodies which are found outside the nucleus. Some of these are called mitochondria, and their task is to break down nutrients and release the energy locked inside them. What we are dealing with here, so to speak, are the cells and the body's energy factory. The remarkable thing is that the mitochondriae are not controlled from the genome of the nucleus; they have their own DNA, and are transferred from individual to individual through the egg. Neither the DNA of the nucleus nor the fathers, therefore, have influence on this important function. One and the same cell thus contains two sets of genomes, the DNA of the nucleus and the mitochondrial DNA. This situation can be explained by the fact that once, back in the dawn of time, two different cells combined in this outstanding collaboration. The formation of a human stem cell, for instance, probably came about through a single-cell organism having permanently enclosed another, and this process happened several times. Seen from an overall perspective the life process is reinforced

Tanzania: crocs and more crocs.

without a petty-minded attitude to the interests of the individual organisms.

Before the organisms could develop into higher levels they had both to invent division of labour and sexual reproduction. Division of labour is usually called cell differentiation, which means that all of one's cells develop differently and take care of different functions. The first fertilised human cell, the stem cell, resembles in its original form many single-celled animals. Then it divides, and for the first couple of divisions the new cells appear identical, but at a certain point the individual cells change and become liver cells, skin cells, bone cells and many others. We do not know precisely why this happens, but we can reckon on the instruction being in the genome, and that the cells are capable of communicating with each other and respond to the signals they receive.

The same process occurs in all higher organisms. Where the human being creates blood vessels, gullet and intestines the plants create a system of tubing which permits the transport of water, nutrients and signal substances from the roots to the top shoot, and the converse. The simple single-celled organisms have thus undergone an evolution, in that the basic model – the cell – has in fact been preserved right up until our day, when the latest models with more RAM and larger hard disks connected into gigantic networks, which have names like human being, gorilla and elephant.

The division of a cell results in a new individual and means that one cell or one single-celled organism has become two. The price is relatively high since the mother cell uses half her energy to produce a copy. If, on the other hand, the cell can be satisfied with breeding half-finished copies or offspring, there is energy to spare and production can be stepped up. With this we approach, paradoxically, the chicken and egg question. Which came first? However, this is not so paradoxical if we make a swift detour via some of our own age's primitive species from the animal kingdom.

The sponges, some of which form colonies like corals do, contribute a piece to the solution of the explanatory puzzle. Sponges can, like the cycads, be regarded as living fossils, since they are found as billion year-old fossils in a fairly unchanged form. They send their immature offspring out to sea. We call them larvae, but they cannot imbibe nourishment, unlike the larva stages of higher animals, which are often known as gluttons. They have a lunch-pack from mother and, starting with this, they must develop into the adult organism which can imbibe nourishment itself and bring its offspring into the world. This gives the mother organism the opportunity to produce more offspring without herself having to drag around offspring as deadweight.

We call this lunch-pack the yolk or the yolk sac and now we're getting to the heart of the matter. The energy supply with which the primitive organisms provide their offspring is what later becomes eggs. First the cells divide and thereby lose 50 per cent of their own mass. Later they will have to content themselves with sending out a copy of their genes with a portion of nutrients.

The yolk developed parallel with the various species and became more and more sophisticated. Naturally, the forces of evolution determined the development of the egg, based on how the offspring could be best protected. There are no rational grounds for immature organisms using energy to move around while they are developing. For this reason reptiles developed eggs which could be laid at an earlier stage of the offspring's development. The eggs took care of themselves and the parents saved energy and could use some of the stored energy to lay more eggs and provide the eggs with so much energy that they hatched as fully developed individuals. Thus there were also savings on care of the young.

Birds followed suit with the same concept, but with the enormous consumption of energy required for flying it became difficult to provide the egg with sufficient energy to produce offspring ready to fly. Birds, therefore, had to take another turn with care of the young after hatching. However, birds have developed a considerable range of solutions to these problems, and most flightless or earthbound birds (e.g. ostriches, ducks and gallinaceous birds) which, as youngsters, do not need to expend energy on flying, have offspring which are so developed that they can promptly follow their parents.

Finally mammals came into existence with a new combined strategy. They retain the fertilised eggs for a long time inside the body where they develop. In this way mammals can move around more easily without having to look after the nest, but, on the other hand, like the birds, they have to make up for this with intense care of the young. On the way there was a brief interlude with marsupials which keep the fertilised egg in the body in the initial stage, after which the small helpless embryo creeps into the specially developed pouch for further development.

The chicken and the egg developed parallel with each other. This is the only natural and logical explanation which, moreover, is compatible with most things in nature. It is, furthermore, quite remarkable that the primitive organisms, such as sponges and corals have located themselves between the strategies of the plant and animal kingdoms. Their larvae swim around freely like animal cells but as adults they remain stationary like plants. Evolution does not necessarily write off tried and true principles but supplements them with new ones.

The ascending level of organisation in nature could give us the impression that we – the human being – represent a form of evolutionary end-product – or merely the highest level – but the human being is, considered in isolation, no more complex than other mammals. The difference lies in our method of organising ourselves. The communicative and ethical putty which binds human beings together is extremely complex even if one can discuss how far this is an advantage or particularly refined. For the time being we are pretty successful when it's a matter of increasing the number of the species and of dominating all others, but for how long? It might only be possible to talk about success if we had managed to lay a long-lasting foundation for humanity. And we've hardly managed that!

The mysteries of life are quite certain

One cool but sunny day in June I'm running up through the steep Yellow Wood Ravine of the Waterberg with a group of crazy joggers who are dead set on running a marathon among Africa's wildlife. Halfway through the ravine I catch up with a young lad who is just as tired as I am and we decide to take the whole thing easy, and look out for small klipspringer antelopes, mambas and leopards, for which this ravine is notorious. All that shows up is the first-named klipspringer plus the tracks of a leopard across the road. Naturally I cannot refrain from talking about the cycas, protea and early life, when he suddenly gets the urge to contribute to the conversation with an overwhelming mass of information about the fact the one can easily create life. Perhaps it is artificial life which is easier to make. Life is produced time after time in sterile bell-jars. It will always emerge by itself. That has been scientifically demonstrated.

I sigh briefly: it's a long way to the finishing line. No, no and again no,

It would seem the tortoise population in Aldabra is growing.

nobody has ever existed who has created life. And, moreover, it is not science. My pleasant fellow-jogger gets the full treatment which I have dispensed innumerable times, while I marvel how he managed to pick up this rubbish. I ask him, directly and sternly, if he believes in God. Like so many other young Europeans, he reacts by being embarrassingly unsettled at this inquiry penetrating into the unexplained private sphere. Now, he doesn't really believe in God and here we're getting to the root of this nonsense. If God didn't create life, it must have come into existence by itself. One collects and renders scientific chance snippets of information and stories and makes one's own mythology, which is 'neither fish nor flesh nor good red herring'. There's no call for blather whether one believes in God or not.

The early single-celled proto-bacteria represent the first expression of life of which we know. We call them primitive but when we measure them against the complicated organisms of our age where billions and billions of cells coordinate each other's activities they do have the most complicated process in common: life. We are talking about life and living organisms, when a collection of atoms is capable of breathing and reproducing itself. Breathing, or respiration, as it is also called, is the energy generator of life and is based primarily on the breaking down of organic material which is produced by the photosynthesis of plants (even plants breathe). There are, however, other forms of breathing, such as the previously mentioned sulphur breathing which was precisely the respiration of the early organisms. Energy is necessary.

The Earth and the universe around us consist principally of inorganic atoms in fairly simple compounds, but at a certain point in time these compounds developed into gigantic organic molecules which became alive. We do not have even the slightest idea how this occurred. In every respect the distance between the inorganic world and life is infinitely longer than the distance between the mother of all stem cells and the human being, a gap which grows shorter and shorter with every day that passes.

The proto-bacteria represent the level of complexity with which biology begins. Biology is the study of life and is concerned with the first form of organised life and its subsequent evolution, from cell to human being. Biology offers no thesis as to the origin of life, or the time before life, but many biologists have attempted to follow the line even further back: from Louis Pasteur and his glass flasks of sterile bouillon – of which my young friend had apparently heard – to genetic researchers in ultramodern laboratories, but so far all attempts have, to put it mildly, been fruitless, and all theories and explanations continue to remain under the domain of religion. Life is still a mystery.

There is no scientific theory about the emergence of life, only ordinary theories, one of which can be read in the Old Testament. This is due to the fact that the human being itself has set certain boundaries to when something is called science. Science, here in the sense of natural science(s), is actually something quite specific and well defined, which, in contrast to current popular conceptions, does not represent the ultimate truth about the world, but the truth of how the human being perceives the world. That is the only thing about which we *can* have knowledge.

Science is created by the human being as a form of cognition which is based, first and foremost, on what we can sense, or, in the popular speech, what we can measure and weigh, as opposed to belief and speculation. Both Descartes and Plato believed that the human brain has the capacity for unlimited logical deduction, just as the new Protestant movement in the sixteenth century believed that everything was there to be read in the Bible. Both the Bible and the human brain have, as acknowledged, their limitations, and gradually as it has dawned on us that the Earth is not flat, that it revolves round the sun and not the other way around, and that the moon does not consist of green cheese; we are simply forced to invent science: a new way of apprehending the world which is not being incessantly contradicted by the senses.

I believe I've covered pretty much everything, but breathing is not what it was and the young runner steps up the pace with this valediction: "Interesting!" Fortunately I've just caught up with an old friend with whom I team up on the way to the finishing line, while a pair of ostriches watches us from the side, running along with us the next hundred metres. Later that night at a so-called after-party I meet my young jogging mate, who has been thinking things over and now wants to know if it is my opinion that, in time, science will solve the riddles of life, because there is a rational explanation, or whether everything is created by God, and therefore cannot be explained rationally. Bit of a mouthful. I do not know whether this is an 'either/or' question but this answer does not meet with his approval. I state, or recapitulate, that science does not have an answer to everything.

Newton is regarded as one of history's great men of science, whose intuitive explanations of natural phenomena have since been proved. He provided, for example, a good explanation for the movements and orbits of the planets. The first he explained with the law of motion, which, in its popular version, asserts that everything in motion continues in motion forever if there is no resistance (air, for example). The orbits of the planets he explained by means of gravity. The Earth and the Moon exert a pull on each other, and the Sun also pulls, which is fortunate, otherwise we'd end up on a collision course. Today we know that this is how it fits together, seen through scientific eyes. However, Newton failed to answer a tricky question: who set it all in motion? Newton's only answer was: the finger of God. How far he himself believed this answer is anyone's guess, but the point is, he didn't have a better one.

> Neither do we have one to the question of the riddle of life
> And it only remains to add that the last answer was actually accepted.
> ... whatsoever King may reign,
> I will be the Vicar of Bray, Sir!

Waterberg is an island in the African continent. The first Voortrekkers settled here a good 170 years ago as voluntary refugees from the British system in the Cape, and attempted to beat life-giving crops out of the dry earth. They cultivated the flat expanses of the highland right below the steepest cliffs. Fortunately the flora and fauna remained untouched in the most inaccessible ravines and ridges of the rocky landscape.

Before me a dammed-up lake bears witness to the human being's struggles to create its own conditions of existence. Right now the lake is full of water, raindrops bashing down hard as the catfish skim along the surface, their jaws, as in many variants of this breed, devoid of incisors, agape to catch falling insects. This is not their daily diet, but it is acceptable when occasion presents itself. Without water, no life, if I may be permitted to repeat myself, but it can hardly be said too often. The last crops were harvested years before, since agriculture has been abandoned in this dry, iron-rich red soil, which now accepts the old trees and flowers, which, in years gone by, hid away in the ravines and shadowy slopes of the mountains.

Waterberg is on its way back to nature, where tourists and UNESCO have already got into the act with titles worthy of preservation to the reclaimed nature without Bushmen or refugees from a hundred years of lonely colonial war. On the other hand, Africa's familiar silhouettes of elephant, rhino, buffalo, lion and leopard are seen at sunset. We are giving the Earth back to nature. But which nature are we giving back to whom? What is original? Can one talk about something original with human absence – is there any sense in that?

It foams at the edge of the dammed lake: conspicuous traces of proteins, metabolic processes and, not least, bacteria. This is obviously a case of pollution, as there are masses of bacteria. The bacteria themselves would probably think otherwise, if they were actually capable of thought. A closer look into the small inlets reveals the origin of the scum: a number of small blobs of slimy substances, some of which are as clear as glass and almost imperceptible, while others are dark and unpleasant.

Slimy bacteria deposits can be found everywhere, in lakes, water courses, sulphur springs, water pipes, human bodies, the forest floor and beneath the ice of the Arctic. Everywhere microorganisms have a tendency to gather in large, complex colonies, which are regarded as filthy accumulations of bacteria. These colonies may consist of many microorganisms, such as bacteria, algae (if there is light) and other single-celled animals, as well as, occasionally, fungal organisms. Some consist of a single organism, but most have an extraordinarily complex composition. A portion of minestrone, please! Even in running water it is possible to find these colonies, where individual species with particularly developed characteristics first colonise the area by placing a hook in the surface of the water. Thereafter other species arrive and attach themselves to the first ones. These micro-universes may be regarded as limited ecosystems in controlled interchange with the surroundings.

A slimy surface serves to protect against attack from outside by chemicals, antibiotics and possibly the immune systems of the host organisms. Biologists call this matter a biofilm. If one imagines a sunken Atlantis with a large glass cheese-dish cover to protect the inhabitants, one would not be far wrong. The organisms inside the cover simply build the slimy cover – the biofilm – bit by bit, and improve it in the process. Various colonies produce various (kinds of) biofilm(s) which can vary according to their species composition, and the surrounding environment. If the ecological circumstances change, the colony responds by reinforcing the construction. The surrounding milieu determines the development inside 'Atlantis'.

If the biofilm is destroyed so that the inhabitants are dispersed they will immediately resume life as solitary inhabitants of the planet and look after themselves. Thus the colonies consist of quite individual and often quite different microorganisms which manage to combine in an organised society in which every cell has its precisely determined function.

When the first inhabitants have reached a certain number the construction begins of the cheese-dish cover (or biofilm), which in technical parlance is called the extracellular polymeric substance. This is a complicated process involving many different genes, both for the formation of the film and to influence the process. It is probable that more than fifty different genes take part in the processes by which the film is formed, developed and controlled. Gradually, as the colony gains new inhabitants, the film is regulated. It is made larger, possibly thicker or more resistant to threats from the surrounding environment.

Various cells and various species each contribute with their particular part of the process. The cells work in the literal sense together on a common endeavour, which, in the process, becomes more and more complex and involves new genes. The project starts from the bottom, and the organisms must communicate on everything. When one cell contributes with the development of a material for the project, it is, at the same time, a signal to its neighbour for the latter to contribute in its turn. Since the microorganisms do not build biofilm outside the colony the crucial factor must be genes which are either activated for this specific circumstance, or can be recycled in other, new functions.

A cell, therefore, is affected by the signal substances of other cells. Its genes encode a protein – that is, a signal substance – which penetrates into another cell, where it opens and closes the genes of the cell in question. The set-up is highly reminiscent of the process which occurs in the human body when our cells are differentiated. Our identical cells cooperate in developing themselves into different cells; in building a human being. For the remainder of their existence the cells in human biofilm communicate as regards maintaining and caring for the system in the best ways. The biggest difference between biofilm and human being is that human cells cannot immediately return to the basic cell function as a stem cell, once they have been converted into, say, liver cells.

As mentioned the biofilm matures with time. It becomes stronger and the inhabitants elaborate the cooperation which now reaches higher levels of complexity. The colony lives off the surrounding environment while simultaneously protecting itself from the latter's dangers. Besides the physical defence and the protection extended by the dish cover the colony also benefits from the nutrients present with less internecine rivalry. They exploit each other's waste products instead of fighting over the same food. This is probably also an exchange of characteristics (genes) such as resistance to, and protection from, poisons in the environment. Examples are known of organisms in a biofilm shutting down all activities when they are at the mercy of antibiotics and remaining in a state of hibernation until the danger has passed.

The price the organisms have to pay to be members of the society is limited growth compared to being free-living individuals. They cannot develop

Lilac-breasted rollers.

recklessly but submit to the rules governing society and only produce offspring when there is room for them. On the other hand, they develop the ability to colonise vacant niches, such as running water in rivers, tubes and bodies where the individual organisms cannot survive on their own.

The ability to organise into complex structures or society with reciprocal dependency is thus not exclusive to human beings, but is apparently an in-born tendency in even the most primitive organisms. Individually they thrive outstandingly in their respective environments, but when the opportunity arrives they act together.

Quite what mechanism makes microorganisms relinquish the free life and submit themselves to inclusion in a system is something we don't know much about, apart from the fact that it is the response to numerous reciprocal signals. The curious thing is that the society consists of different species: does this in its turn then mean quite different genes or gene pools?

Penicillin-resistant coli bacteria produce a signal substance which warns other bacteria, but in the attempt to save 'the family' it risks perishing itself, because they use enormous amounts of energy to produce the signal substances. This does, however, make good sense, because the nearest bacteria are copies or very closely related 'variants'. The bacteria, so to speak, protect their own gene pool, completely in the spirit of the theory of evolution. The altruism here is absolutely understandable, but how should we understand the biofilm? Is the mechanism in this context the same as the reciprocal altruism we observe in the bat, even though micro-organisms hardly go in for mutual grooming? How is mutuality ensured if it isn't pure chemistry?

The $60,000 philosophical question is, how far can this universe of mutually dependent organisms be described as the sum of all the individuals? Or is it greater than the sum? For the moment we must simply state that they communicate with each other and adapt to each other in a hair-fine system where everyone knows their place. A society for the greatest benefit of all, where everyone toes the line laid down by the community and the one who is found wanting perishes and is swallowed up. The price of ultimate survival under difficult conditions is loss of independence and one is obliged to have recourse to the same adaptability as the famous vicar faced with all kinds of liturgy, Protestant, Catholic or whatever.

Social insects with fascist tendencies

Encephalartos eugène-maraisii is the cycad palm's scientific name and the only reason I dare to name this admirable tree yet again is to embark on a discussion of South Africa's great naturalist Eugène Marais (1871–1936) who gave it his name. For years Marais lived in the barely accessible Waterberg Mountains where he also collected the first examples of this peculiar plant while studying pretty much everything else nature had to offer. He was, rather, a man of many talents, studying and concerning himself with such widely differing disciplines as medicine, law, biology, politics, literature and journalism.

Following the death of his wife in childbirth he became something of a loner, original and opium addict though without losing his ability to write outstanding articles and poetry and carrying out unique behavioural studies in the wild. He is probably among the very first to undertake actual systematic natural scientific studies in Africa. Among Marais' favourite objects of study were baboons and termites, which he described in detail in the books *The Soul of the Ape* and *The Soul of the White Ant*. Both books are filled with incidental stories of birds, snakes and everything else which entered his immediate surroundings. He was the incarnation of a classic field biologist who studies and searches for explanations for, so to speak, everything he finds along his way. During intervals in his investigation of, for instance, the hunting behaviour of the baboons, he studied the honey-guide or grazing antelopes.

His study of the life of the termites in the dense, enclosed mounds was a project on which he spent ten years, and, moreover, continued until the day he died. He dug down deeply into the nests, observed the insects, dissected them under the microscope and conducted a mass of experiments. At night he lay on top of the termite nests, shining his torch light down into them in order to discover the secrets of these curious insects.

The initial results were a series of articles in the South African media and later the book mentioned above, in which he puts forward the epoch-making thesis that social insects such as termites should not be regarded as single individuals but the whole colony should be seen as a giant cohesive organisation, in which every function is carried out by specialised individuals with a total system of division of labour. The superorganism, as it is currently expressed, is characterised by the fact that it functions only in its totality.

As a researcher Marais was highly untraditional and holistically oriented, unlike the contemporary European natural scientist who was reductionist. He wanted to know how the whole colony functioned. For Marais it was not sufficient to measure how many grams of cellulose a colony metabolised per day. Several contemporary scientists had got hold of the idea of the complete, well-organised insect society working from the study of ants and bees, and the term superorganism probably emerged for the first time in 1911, when it was introduced by the insect researcher William Morton Wheeler of Harvard University. Marais interpreted the idea quite literally and described the specialisation of the termites as analogous with the organs of a human being, and thus the queen became the brain, and the walls constructed around her the cranium of the organism.

The discussion of the analogy between people, society and superorganisms was pursued hotly in the interwar period and enjoys rude health to this day. Marais believed that social organisations like termites and higher animals, such as baboons, do not react as individuals, but wholly or partly in conformity with the contemporary holistic philosophy, which is, to put it briefly, that the whole is greater than the sum of its parts. This may sound a bit hippyfied, but nevertheless the notion is discussed again these days having re-emerged from many years in the darkness.

I sit with Marais' termite book on the spot where he made most of his observations, wolfing down the 144 pages with a passion. I'm not really into reading natural science with a century behind it in an age when every book of that kind is obsolete before it is even delivered to the printers, so this is an exception. I read with mounting astonishment at the fact that Marais' books have been met with so much criticism – and they still are – from well-established scientists who will not accept Marais as a scientist because he does not remain within the parameters of our age's exposition of research results. He consistently pursues the holistic theory and may come across as blind to other possibilities, but what is most provoking is that he insists on talking about a whole organism with brain, spine and bloodstream when it is, after all, just a mass of small individual insects.

The landscape around me contains numerous termite mounds of various heights and maturity, but the diversity of the various species makes a definition based purely on the nature of the colony impossible. It is the rainy season and new colonies are being started everywhere. The new mini-dwellings are seen as small fragile tubes sticking up out of the ground and indicate that here a new colony is being created. In many places they stand in a cluster because many have settled there in the same place. The nests are not yet sealed and the hard-working insects live a dangerous existence, since, generally speaking, all living things around them regard their protein-rich bodies as tasty titbits.

This is also true to a considerable degree of human beings who are willing to go a long way in search of a delicious meal. The savannah's other mammals, reptiles and birds wait for the first rain and flock threateningly around the mounds where they await the season's termite swarms which are all released on a single occasion. Millions of new termites fly out in an assembled mass and the species obviously operates on the assumption that only a few avoid ending up in the stomach of one animal or another. The wings are nothing special, serving roughly the same purpose as the pappus of the dandelion: to spread as many as possible with the wind. Often the flight is only a few metres in length, but immediately after landing the insects bite off the unusable wings, and the males signal to potential partners by stamping on the ground. When the female hears or feels his mighty pounding through the earth she seeks him out, after which they form a pair and immediately dig themselves into the protective earth.

Contrary to the current notion termites are not directly related to the other social insects such as bees, wasps and ants. Their common ancestor lived about 250 million years ago, when our cycad palm was in charge – whoops, there it goes again – when the species was split into one group which later became bees and ants, and another group which developed into cockroaches. It was only a hundred million years later that the latter sent out an offshoot with the future termites. Society-founding insects have, accordingly, developed from two different points of departure and yet have arrived at a fairly uniform result.

How it came off we can only boggle over. There are various ways which I went into earlier. One possibility is homology. Here the same characteristics are drawn from the genome and adapted for current use. The slumbering proto-genes are revitalised and are expressed when environmental factors develop them. One might harbour the suspicion that the social systems are so complex and incorporate so many different genes, that the method is not quite obvious. Perhaps the regulator genes with the governing functions play a leading role? We develop whatever there a use is for; out of whatever elements are to hand. The easiest thing would be to imagine a combination, in which case we would find repeated, but not completely identical, regulatory mechanisms in the two groups of social insects. In any event we can state calmly that the organisms' possibilities for adapting to both conceivable and inconceivable situations are somewhere near endless, and that, accordingly, life is capable of percolating down into all the cavities.

In the first chapter Marais describes his observations of the termite queen in a fully developed nest. He asserts that the queen cannot move, that she is shut inside a cell from which she cannot exit, that she is too large to be trans-ported, and yet she still turns up in another cell. One may therefore conclude that there must be several queens 'developing', and that when the functioning queen becomes too large and has served her time, she is killed and eaten. This conclusion is quite incorrect, maintains Marais. The analogy with bees simply doesn't apply since Marais has marked his queen and can state that it is the same queen, who continues to rule year in, year out in ever new, larger cells.

In the final chapter Marais returns to the queen. A house in Pretoria is completely surrounded by termites which threaten the total destruction of the human dwelling. Marais offers his help in return for being granted a few days in which to study the termite colony at close range.

The beginnings of a termite mound.

Through his experience he finds and exposes the queen's dwelling down under a dark corner in the house. For several days he studies the way straight ranks of termites come in via the entrance to the queen round the clock and perform some quite specific functions, after which they leave via the back door. One type of termite delivers a drop of nutrient to the queen, another fetches an egg, while a third administers a curious massage of the queen's gigantic body and disappears with a liquid – probably exuded by the queen which is used to feed the small newly-hatched termites. At one point a lump of earth falls on the queen, and after a few moments all the activity in the colony is halted; this even applies to activities far from the queen. A rescue action is mounted: the queen shakes her head rapidly and is subsequently ready to continue, after which the routines are resumed.

Later on Marais removes the queen and the entire infrastructure of the colony breaks down instantaneously. Marais concludes that the queen is the brain of the colony, which controls all signals and work functions: without the brain, no life.

Marais was never accorded recognition for his epoch-making research, because he was a Boer and naturally anti-British during the Boer War and in the years that followed. Like so many of the other defeated Boers he was so bitter about the English that he insisted on writing in his local language Afrikaans, which is a development of the language spoken by the Dutch settlers. Even so, Marais became known beyond his narrow, local public because of an unpleasant plagiarism case in 1926. The world-famous author and Nobel Prize-winner Maurice Maeterlinck had himself never seen a termite and had to base his work on the observations of other people. In addition he put forward some theories concerning superorganisms which he had read in Marais, whose work contains so many original thoughts and observations. These theories Maeterlinck appropriated and passed off as his own. The plagiarism was discovered and Maeterlinck's posthumous reputation must live with this unnecessarily incurred blemish.

In the interwar period Communism, Fascism, Nazism and other totalitarian thought roared out of the depths and Europe's intellectuals threw themselves over works about systems of society, including analogous descriptions of well-organised insect societies. Both Maeterlinck and Marais managed profoundly to catch the spirit of the times and, had Marais chosen to write in his first mother tongue, English, history might well have been different.

Even in Denmark well-educated politicians, artists and intellectuals permitted themselves to discuss the advantages and disadvantages of democracy and other forms of government, in which the governing institutions were not popularly elected but where the state was led by one dynamic man with a high societal morality. Even democrats were agreed that something as serious as the country's welfare could not be left to just anybody: women, for example.

Plato's *The Republic* was diligently perused. In this work Plato has Socrates develop the model of an ideal society with a perfect division of labour, in which every citizen does what he is best at (the female sex doesn't really feature). Citizens of the society who are good at tilling the fields should naturally concentrate on that activity, just as builders, carpenters and smiths should be solely concerned with those worthy occupations. We are talking here about the rational, reasonable society where everyone knows his place. The society should naturally be governed by a Guardian and a Guardian Class, which is, of course, best suited for purpose. The most gifted are singled out in childhood and should concentrate throughout their entire lives on leading and governing.

For Socrates it was understood that his excellent system would exclude the politicians who, at that point in time, were actually at the head of the Athenian democracy. The leading men were, on the other hand, not as enthusiastic about

Socrates' unostentatious outpourings, choosing to requite his endeavours with a goblet of hemlock.

The description of the life of social insects comes pretty close to Plato and adds a dimension to the argumentation about well-ordered society ruled from the top: yes, but ... when Nature herself ... the division of labour among social insects is almost perfect and extremely specialised. Each knows its place in the formation of society and wouldn't dream of deviating, a course of action which would lead moreover to instant death. Everyone works for society for the greatest common good and gladly sacrifices itself to the greater whole if necessary.

The tendency of social insects to sacrifice themselves – the cultivated, pure altruism – was a problem for Darwin and many early adherents of evolution, because the insects' massive self-sacrifice threw a spanner into the works of natural selection. If half of humanity commits suicide, how can one ensure that it is the fittest (best-qualified) who survive? Natural selection is the ultimate paradigm which is not susceptible to comparison. Either it operates everywhere and all the time or it doesn't.

Darwin brooded greatly over this threat to his theory. There had to be a natural explanation. There was: but it only emerged many years after his death, when people began to regard single individuals as vessels to transport their own genes. The human being, the termite, the donkey, the budgerigar are constructions which various genetic families put together to transport themselves farther along the path of evolutionary development. Neither a human being nor a seal is the result of its own will, but of hereditary dispositions: the genes.

One genetic family builds people while others build seals. The single genetic family contains sub-families which build more or less uniform individuals, and the best-constructed vessels are accordingly granted the right to sail farther towards the spring tide of development. This image makes it easier to understand the altruism of social insects, because all of those in the same colony are closely related and transport the same family of genes. The principle is that he who helps his neighbour is, in reality, promoting his own genes, even if the vessel of the individual in question goes down. This is the same as what happens with the aforementioned coli bacteria and probably also with all other genuine altruists.

Theoretically, the optimal goal for the human being is to produce a clone of itself. This is, as is well known, not (yet) possible, but we or our genetic family can achieve (almost) the same result with half-shares in two children or four grandchildren. If we have three children our genes are represented by three half-sets, which is (a) surplus of our own specific genetic family, even if the genes are mixed in mating.

As a human being one may feel a certain reluctance to being controlled by one's genetic family, which, after all, opens up the possibility for a mass of discussions: a corner of the veil which covers them, I have already lifted. We can, however, console ourselves with the fact that there is nobody who knows how these things hang together, and all the world's greatest philosophers have given themselves grey hairs in speculating on the problem. The philosopher Kant was already aware of the question of free will in a world which was dominated by the system of nature. He produced an elegant solution to this problematic by dividing our approach to the world into two categories, which consisted of the in-born (a priori) and the acquired, which we call experience (a posteriori). We come into the world equipped with certain preconditions which today he would have called genetic, and which contain the structure, with which we order and categorise experiences or science. The structures also supply us with moral law (a priori) which makes us capable of acting freely, in that we can follow reason, which is to give and act in a manner that allows us 'to will' as Kant expresses it.

Superorganisms and swarm intelligence

From my lookout in the Waterberg I can see, at the same time, biofilm, sugar bushes, termite colonies and a few cycads. There are of course many other plants and birds in the neighbourhood as well as an occasional mammal which crosses my field of vision. It becomes ever clearer to me that nature, the ecosystem, the niche, the species and the organism are, in more than one sense, connected vessels. All the organisms exist in a close relationship of interdependence, often so strong that one species determines another's existence. Where does one end and the next begin? Neither ecosystems nor organisms have (firmly) fixed borders with each other.

What, in basic terms, constitutes an organism? On the face of it this sounds like a foolish question, but it demands, nevertheless, a complicated exploration if asked by an extra-terrestrial who has just landed on Earth. On the first encounter with a human being the alien would probably be in serious doubt as to what kind of organism this is: a movable lump of organic material consisting principally of several hundred billions and billions of bacteria, which secrete many hundred billion dead individuals every day. To whom should one address oneself to if one wishes to establish contact? Is there a helmsman?

Human beings are first and foremost human beings because we are in agreement in defining each other as such and our organism consists of precisely the current assembly of cells which are wholly or partially coordinated by our brain and central nervous system. For other billions of organisms, namely the single-celled organisms inside us, we are merely a random living space, an ecosystem which they fill to bursting point. Thousands of species spread all

over the surface of the body and not least the most moist depressions, into the passages of the ears, in the corner of the eyes and in the nose, mouth and throat and hence around the body. Mutually they hold each other in check, some take over in certain periods when others go on the defensive.

We thrive superbly in each other's company. The human being is completely dependent on his so-called natural bacterial flora, which, on its side, is dependent on the human being. Every person has their individual bacterial flora, which is determined by choice of food, drinking habits, climate, clothes, the way they wash and sweat, and certainly a considerable number of other factors. Once in a while there is disorder in the system, the immune defence system fails and the bacteria penetrate into areas from which it were better they were kept away. We become ill. Most of the bacterial illnesses are not due to massive hostile attack, but our own, benign bacteria which have lost their way.

In its way the human being and its bacterial flora is a superorganism, for without each other we cannot exist, which is precisely the difference between us, termites, bees, ants and similar social societies on the one hand and biofilm on the other. Only the last-named consists of free, individual organisms, which for a time have come together in a common project, but which quickly and simply can return to the free individual life.

Termites, for example, are locked into their places in the system. They are born as highly specialised, and can only perform the simple function for which they are designated, and they are completely dependent on all the others also accomplishing their tasks with great precision. It is a system with no room for fluctuations. Consistent with these purposes is a compatible dwelling, which is constructed with extreme precision. The termite nest consists of a labyrinth of passages, the walls of which are cast in a hard, specially produced cement. They can be constructed so thin that atmospheric air from outside, and the colony's own kinds of gas from inside can pass through without their appearing any visible openings. This curious system enables the colony to breathe and regulate temperature very precisely. It is difficult to view the colony as anything other than a large assembled organism, which both reproduces and breathes, which is an exact definition of life.

It is thus also easy to pick up on the analogy between termite colonies and other organisms, not least between the termites and other organisms' cell differentiation, e.g. our own. The cells which come into being through the divisions of a fertilised human egg, are only uniform through a few cell divisions, subsequently they specialise into, for example, skin and liver cells. When a cell has specialised once, it is fixed in its place in the system. A liver cell is a liver cell, just as a brain cell is a brain cell, regardless of the fact that every cell's DNA contains all the functions and theoretically can develop into every other kind of cell. Once cells have specialised, other potential characteristics are shut down forever, something we work hard to avoid, however, in stem-cell laboratories.

In biofilm, in principle, things function the same as in the human being, for here the cells also differentiate, but at the same time they retain the ability to change back again. Even though biofilm is a brilliant response to a quite special need to adapt, it can actually, in superb fashion, mark a step towards a higher complexity which is represented in cell differentiation in higher organisms.

The termite colony resembles far more closely a fully developed organism, where the individuals correspond to the cells, or rather the organs in a larger organism, such as the human being. Even the nest is inseparably bound to the termites, just as the human being is connected to its skin. From this point there are two paths within biology. One states that it is a very exciting or funny analogy, nothing more nor less. The other route begins to philosophise about life, organisms, intelligence and higher powers.

As mentioned above, termites are at least hundreds of million years younger than their distant cousins in the ant and bee family. They entered the world arena in order to occupy a vacant niche, which fell available by virtue of the many new cellulose-containing growths. Generally speaking wood can only be broken down by bacteria, fungus and single-celled animals. Only they produce the necessary enzyme, cellulase, which can cut cellulose into digestible mouthfuls. Higher organisms are simply compelled to ally themselves with the micro-organisms in order to secure a share in the enormous quantity of food which is tied up in many plants. Not even the termites have developed the ability to break down cellulose, but fortunately they can rely on various symbiotic joint ventures with micro-organisms.

A large group of termites (Macrotermes) have, for instance, developed a complex form of agriculture which is highly reminiscent of the human variety, and is somewhat more complex than the nature management of the elephant. In the base of the dwelling they simply cultivate microfungi which take care of the heavy work of metabolising – and at the same time the increasing of – the nutrient content in the cellulose by three to four times. First the termites chew and soften the wood mass, after which this is built up into vertical walls which are inoculated with the fungi. The fungi grow well on the mass which is broken down and reconstituted into digestible materials. Both the metabolic waste and the fungi themselves. The fungi do not, however, come into existence of their own will. Therefore the year's new generation of termites is provided with fungal spores before they leave the 'home' to move and form a new colony.

Like the termite colony, most social insects accomplish many complicated tasks. Superorganisms manage to act as a single organism with clearly defined reaction patterns, that occur through extensive communication, which can be based on sounds, odours, and behavioural patterns. To put it briefly, all senses can be in use. Joint action of many small brains is something which today we call swarm intelligence. The individuals can be compared to the brain's individual cells, neurons, which are in contact with each other through the synapses. Taken as individuals the neurons are probably not particularly bright – but together they create an incomparable unity. Insects, taken individually, are not too bright, but when working together they are a completely different proposition.

The expression 'swarm intelligence' is used more broadly, e.g. with the age's electronic, social network, which becomes stronger and more significant every day. Perhaps Facebook and corresponding networks represent a new form of swarm intelligence among human beings, which make large groups move or react as an assembled body

Marais, who was mentioned in the preceding passages, performed a relatively basic experiment with a termite colony. Quite simply he drove a steel plate down through the middle of the colony and was able to state that the termites worked in a fully coordinated fashion on both sides of the steel plate, which should prevent any form of communication. From both sides the insects built bridges over the steel plate and they made contact with each other with millimetre-fine precision. This caused Marais to conclude that the termites were capable of more in mutual coordination than when acting alone: that something existed which today one would call swarm intelligence. These are exactly the thoughts which today are put forward concerning new technology-based social networks and experiments with artificial intelligence. As the Scots put it: "Mony a mickle maks a muckle."

The big question is: what makes single-celled organisms, social insects and human beings organise themselves into society? Mutual gain? Is it reciprocal altruism, and if so, where does it come from? Is it the same mechanism which is passed down in the genes from the time of creation? As I have already mentioned, it is possible to boil the answer down to two positions, which pervade the scientific world.

The position most widespread among professional biologists continues to be the reductionist one, one representation of which is that of the father of socio-biology, Edward O. Wilson. They believe that everything of fundamental importance is passed down in the genes, and that by observing these we can predict how a given organism will develop. This does not exclude a certain environmental influence, but the principle features have been established, so after all there are no more to look for.

The other position is represented by such people as Rupert Sheldrake, who has taken up the old holistic tradition, as promulgated by Gregory Bateson, Eugène Marais and numerous others. They are agreed that a gathering of individuals always contains more than the sum of its parts. To put it more colloquially, we all have pieces of the whole which only function when the whole assembly is complete. Jung talks about the collective unconscious, and that is precisely the concept which Sheldrake borrows. Not one of us can use our individual fragment, but when we are assembled we can react as a totality, without thus being fully consciousness of how the system functions. While Wilson always wants to explain an actual action with an odour, a pheromone or another specific signal direct from the DNA, Sheldrake will talk about collective memory, which functions both within the single individual and outside.

Maybe we are no further enlightened, but where Sheldrake contains the possibility of a spiritual dimension, Wilson is the hard-nosed materialist where the rational, strictly scientific has taken over the spiritual. Religion is passed down in the genes and we are nothing other than what can be read in matter. Here the matter-of-fact Kant would have said, "Yes, yes, that's all very well, with these genes, but who made them and who provided them with their respective dispositions?"

True faith expands cognition without standing in its way.

Gorilla – a genetic construction

Like eternal snow the mist shrouds the Virunga Mountains in the heart of Africa and nourishes the lush rainforest around the two volcanoes Virunga and Karisimbi. The African legend-landscape is not easy to spot in the impassable wilderness, but if one were to climb high up in the scrub and poke one's head out through the tops of the trees one would experience the unreal sight of a fleeting cloud cover which spreads gently over the dark silhouette of the jungle to dissolve and form again in the slight breeze.

There are neither roads nor paths, and where would they lead to anyway? The jungle is masssively dense. In the absence of a machete-wielding Indiana Jones we have to squeeze and struggle through the undergrowth – often on all fours, rucksack before head so it doesn't get stuck. At regular intervals we hear monkeys signalling and birds calling out warnings, while the broken branches, uprooted trees and crackling thicket tell us that the elephants are doing considerably better than we are. Often we pick up the scent of buffalo, acrid and characteristic, but we don't see them, which is fine by us. A pair of gardening gloves is a good idea when pushing aside and moving the undergrowth. Something is burning and something else biting. Every so often, my twelve-year-old daughter has tears in her eyes, but she has to keep fighting and manages bravely without a murmur. We are a long way from the safe haven of an everyday life with TV series, Saturday sweets and soft drinks. I am quivering with anticipation, though I'm finding it difficult to grasp that I am actually here, that, as a quite normal human being, plagued with a seventies high-interest student loan, I can stand here in the middle of *Animal Planet* hunting for one of the fabled man-apes. Humankind's hairy cousin. The gorillas in the mist.

I don't know much about gorillas – a little about biology which I am attempting to expand – but this particular day was to be the turning point, when human beings, apes, plants, cells, science, philosophy and religion started to come together for me. But I only realised it when thoughts went down on paper for this book many years later.

We share ancestors and many genes with the gorilla of our time. That ancestor lived in the primeval forest of Central Africa, from which a small, pushy, enterprising branch of the family made its way out onto the open savannah. If we go further back in history, the descent can be traced back to primitive apes and 'half apes' which have occasioned a plethora of new 'Missing Link' discoveries. Lemur-like species from Madagascar or ancient fossils like the primitive ape which was found in Germany a few years ago. That ape was, moreover, dubbed Ida and under tremendous media coverage hailed as a sensational discovery of the human being's ancestor, which is perfectly correct in as much as every ape fossil is entitled to the same illustrious appellation. The incorporation of reptiles in the descent would form the fertile breeding ground for a number of sensational revelations, such as Ulrik the Crock or Prince George the Frog, just use your own imagination.

The history of the human being is entangled with the developmental history of all other organisms, and the principles are exactly the same as with amoebae, birds and elephants. Genetic differences between the species can be read like the numbers on a clock, which, with considerable precision, tell us that the species which developed into the human being, separated from the rest of its companions some five to seven million years ago. This number depends on whether the chimpanzee is grouped on the human or the ape side of this borderline.

The human race became geographically separated from its ape ancestors, something that probably happened in a 'sliding sequence'; the remaining group developed into the gorillas of our era. The gorilla has also changed and hardly resembles our common ancestor. They have become larger and fantastically well-adapted to life in the dense jungle, where they have become familiar with all imaginable food sources and the growth cycle of the available plants.

The group which became separated remained together – or developed together – for a further two, three million years, until history repeated itself and a new division took place. The new split resulted in a chimpanzee line with two remaining species, and a human line with one remaining species. If one lists the two remaining groups' genomes side by side, there proves to be a 98.4 per cent coincidence. The difference between human being and chimpanzee is thus only 1.6 per cent, but the remarkable thing is that the difference between both lines and the present-day gorilla is twice as large – around 3.2 per cent. This confirms the course of development outlined above, and makes the human being and the chimpanzee each other's closest relatives.

In genetic terms the chimpanzee is therefore closer to the human being than the gorilla. Even though one should not 'over-interpret' the genetic kinship, the placing of the human being between gorilla and chimpanzee departs from the current notion of the human being as an evolutionary zenith; completely consistent with classic evolutionary teaching, which does not compromise this development.

What does it mean, fundamentally, to say that our genes are very similar? Is it a meaningful statement at all? The human being has long searched for the 'Missing Link', the connection between apes and men, but the palaeontologists of the future may well end up stating that fossilised human bones are the 'Missing Link' between gorillas and chimpanzees. Before we go a step closer to the genetic similarities, we might take a round trip in the history of genetics.

In 1953, Crick and Watson published a model of the complicated DNA molecule and thus brought a peaceful conclusion to a battle of years duration over the structure of genes. Ten years later they were awarded the Nobel Prize for this achievement which was partly based on the research of Rosalind Franklin – but in their haste they forgot to mention her. At last we had the connection between genes and proteins: the definitive model of the genetic structure, the building brick of life. All characteristics are incorporated in the genes, each one of which corresponds to a protein. The individual organism can thus be read as the sum of the genes or the corresponding proteins.

In reality this very simple model originates in the time before we had control over molecules, and goes right back to the end of the nineteenth century when the aforementioned monk Gregor Mendel carried out the famous experiments with pea plants. Mendel demonstrated that the organisms he had investigated inherited characteristics from both the father and the mother through sexual reproduction. The characteristics can be found in dominant, recessive and equal form. Mendel's study of heredity led at the beginning of the twentieth century to the dogmatic notion that all characteristics are hereditary, and thus it is pointless to concern oneself with anything other than looking for these characteristics. The characteristics were given the name genes. Crick and Watson's gene model is a beautiful scientific solution. We call it beautiful because it is highly regular and logical, but, in time, it has become apparent that the mechanism is somewhat more complex and unpredictable than that.

Small genetic differences may have a considerable effect, depending on

Rwanda: Mgahinga gorilla, in a national park.

where they occur. A slight change in a gene can bring about massive alterations in its respective protein. A single mutation can turn a protein into a fatal inherited disease, such as cystic fibrosis.

On the chemical level the difference between men and women boils down to the quantity of the male hormone testosterone. If an embryo produces sufficient testosterone it will develop into a man. Even though two organisms have exactly the same genetic structure, because they may be created by the dividing of, for example, two stem cells, two amoebae or two bacteria, their genes may, under certain circumstances, be expressed quite differently. Large quantities of protein produce an effect quite different from a small dose and, furthermore, it is also apparent that the different cells, in particular brain cells, have various capabilities of binding the proteins (receptors) which can bring about contradictory effects. There is an amazing amount about which we know amazingly little. Therefore one cannot always translate scientific observations

A very young gorilla, in a Rwandan national park.

into logic. We do not know all the interim stages but let us give some of them a closer inspection.

According to the classic notion, our genome, also known as DNA (the name of the molecule) controls the development of the entire organism through the formation of proteins. A single gene contains the code for a single protein. On the face of it, then, a simple model. Some proteins have, however, a greater effect than others. Some proteins have many different functions in different places in the organism. Some proteins attach themselves to the genome and control how much is produced of a second or third protein. A protein can start or shut down processes which involve fifty or possibly even hundreds of other proteins.

What we see emerging here is an extremely complex picture of the proteins which may have one or other types of independence of the genome. Every life which begins in a fertilised egg, receives, as mentioned earlier, a lunch-pack. The lunch-pack contains the mother's proteins, and these enfold the genome and determine what happens next: which genes shall be activated, and to what extent. Accordingly it is not the genome itself which decides, but solely the proteins of the mother. Later the new organism's proteins will take over these functions, but we do not know the significance of the mother proteins' first and original initiation of the processes. One could also ask "Why should the functions of a new life start with the mother's proteins, if this has no significance? Why should there even be interaction between proteins and DNA?"

Logic gives a helping hand: one set of proteins emanates from the genome, which subsequently receives a response in the form of new proteins initiated by the first ones which cause the genome to react differently. If the genome does not react to the proteins which currently enclose it the process is, however, meaningless. A completely unthinkable situation. Something has happened along the way from the moment the genome signals the formation of the protein to when the protein attaches itself to the genome. We can, therefore, conclude that the genome allows itself to be influenced by the environment. Our genome changes expression under the influence of the environmental conditions. Genes, naturally, do not change, but the orders they receive do and they obey.

In front of a gene there sits a small mass of DNA which regulates how much of a specific protein a gene shall produce. This is called a promoter, but it doesn't, itself, determine anything at all. It is exactly the opposite case with the protein which attaches itself to the promoter and determines what will happen. Let us look at an experiment with the gene which produces some of our neurotransmitters, the endorphins. These agreeable substances which were discussed in the sections on plants, we have in common with, among other things, the poppy. They are painkillers, they control our well-being and are of importance for our learning and social adaptation. If one replaces the human being's own promoter of the endorphin gene with the promoter of a chimpanzee, production drops by 20 per cent. The chimpanzee is a smaller animal, so maybe it needs less, maybe the smaller quantity has a greater effect on the chimp. A growing number of 'maybes', but the role of the promoter is important, yet it still does not function without signals from a protein.

The cytoplasm of the cells is filled with small independent sub-units, which bear the general heading of organelles. These include chloroplasts, mitochondria, flagella, ribosomes, vacuoles, etc. each one of which performs vital functions. Some of these come, like mitochondria, from bacteria, which, previously in the history of evolution, were enclosed in the cell and have their own DNA. The whole portion comes, in the first instance, from the mother's egg and thus lives 'its own life' in relation to the cell's genome. Every single human being starts their life with all the good things from a mother cell, including a mass of proteins. But why is this? What is the significance?

The beautiful model with the simple genetic code, which can be explained to every school-kid is indeed still beautiful, but the processes in the organism are dauntingly complex. They have been moving along for millions of years, and every generation has made adjustments. Genes and proteins have been infinitely recycled. Suddenly, for example, we discover a protein in the brain which is otherwise known to control the production of urine in the kidneys of certain mammals. In the brain this protein plays a completely new role as a regulator of social adaptation.

Although our knowledge fills more books today than it did twenty-five years ago, each new discovery returns to the awareness that there is still more we do not understand. For now we must conclude that life is an extremely complicated business which cannot be explained with simple models. It also looks as though we must acknowledge the fact that characteristics such as intelligence, empathy and a sense of justice, which are linked to the personality, are more widespread in nature than we originally believed. They are difficult to incarcerate in molecules, but much suggests that they can be found there, even though we do not know the protein responsible for free will.

The coincidence of genes does not necessarily make apes and humans equal, but neither can it document the opposite, so we must search farther afield for the differences which justify the human being's biblical sovereignty. For want of anything better many scientists have jumped at the conspicuous difference which is called bipedalism, i.e. the ability to walk on two legs. The human being, or the branch we often call humans, is the only ape which has seriously lifted itself up, while our closest relatives still put their knuckles on the ground. Walking upright cuts energy consumption by half and permits a better view around us. There are, thus, good reasons for this change, if one wants to get down from the trees and out into the open spaces.

Bipedalism is definitely a marked difference, even though it can hardly be said that it adds to the concept of humanism a special dimension which justifies the gulf between us and our cousins in the treetops. Our hairy relatives have, though, refined and optimised their living patterns in relation to the surroundings, and this includes their own social relations. They have created the basis for an optimal existence, in which their species can fill the allotted space in the universe – i.e. what we call adaptation. In the preceding period we developed into human beings, but where is the actual dividing line? Most people are of the opinion that the separation of the lines which led the chimpanzees and human beings respectively is the point zero of the human line, but this putting asunder doesn't actually say anything about when the apes of this line became humans or why, does it?

Monkeys in the new World

From the frozen lakes and snow-covered volcanoes of the Altiplano I wandered onward through the beautiful barren Atacama Desert towards the coast of Chile. The sky is more blue than blue but there aren't any monkeys. Even though we find distant relatives in the jungle on the other side of the Andes, they have stayed where they are. They have had no need to conquer new territory, adapt to these relatively inhospitable regions on the dry west coast of the continent. Perhaps it is one of the truly great differences between us humans and the apes. While the apes concentrate on the absolute adaptation in the environment, we spread out in hopes that there may be something better on the other side of the mountain. There is no doubt that we have an inquisitive gene which rewards our brain with agreeable substances when we satisfy this pleasure-producing trait. Are we alone in this respect?

It was in the southern winter of 1982, four years before my first ape safari in Rwanda. Running through the tinder-dry, brown landscape towards the sea was a narrow riverbed, but before the scanty water blended with the cold Humboldt Current, it spread out in a narrow, fan-shaped green oasis which is known today as the frontier town of Arica. This tiny breathing hole in the comfortless, inhospitable mountain landscape was undoubtedly a bridgehead for the human being's conquest of this last continent more than 10,000 years ago. This assumption is supported by the mysterious stone configurations in the hinterland, a landscape of early agriculture and advanced burial chambers. Fertility is richly present, if you can spot it; in the cold sea, rich in oxygen live enormous quantities of fish and seabirds.

Our ancestors had good opportunities to construct a life with surplus and culture when they finally arrived here after a long and difficult journey through Asia, Siberia, over the Bering Strait and down through the American continent. An advanced agricultural, fishing and political system gradually developed over the whole region. A grandiose civilisation was created. Some individuals made it as far as Tierra del Fuego, where Darwin encountered them during the voyage of the *Beagle* and described them in quite dramatic turns of phrase as savages who had no control over their feelings or anything else for that matter. Perhaps they had had bad experiences.

The first immigrants in Las Americas came over the present Bering Strait more than 15,000 years ago. They may have come in several stages, but there is nobody today who knows for certain. On the way they left families which developed into Inuit, Sioux, Mohicans, Navajo, Aztec, Maya, Inca and many other high cultures. In the Amazon they burned the forests and laid out large agricultural fields fertilised with sewage and refuse which mark the whole of the impassable jungle to this day.

A great distance away I see smoke clouds from a vehicle coming rumbling along, and yet again I am lucky, for hereabouts everyone is tremendously friendly. A solitary gringo on the road can always get a lift. You would never be refused. The kindly driver tells me about his good life in Pinochet's country. He is happy and satisfied, life seems to have blessed him with prosperity and a family, and I politely refrain from telling him about my former particularly active participation in protests against the general with the caricature, pompous appearance accentuated by the small ridiculous Hitler moustache. This was about the only amusing thing about him.

Beautiful Chile is still a dictatorship, but this is not something you hear or see much about, even though it is only nine years since the country's civilisation was despoiled by two nearly equally large, irreconcilable wings. A real family feud, where the strongest cousins won. The weak were obliged to crawl in humility and terror took care of the rest.

My driver gets serious, something is crying out to be said; there is one thing more he wishes for in life and that is to kill four – written four – Argentines! This decent paterfamilias suddenly looks scowling and evil. NO. I'm not mistaken. This is not a matter of communists, trade unionists or even social democrats, which would have made some sort of sense, but, instead, absolute bloody Argentines. This is no laughing matter; we're in serious shit here. I never succeeded in finding the exact cause of this curious, even deeply felt desire, nor for the very precise number. I mean, why stop at four of them?

Argentines and Chileans were among the original groups of African emigrants who headed for the Middle East. Subsequently they settled in Europe, but before that they split off from their cousins and wandered eastwards, ending up by creating the aforementioned Pre-Columbian cultures throughout the whole of America. In Spain the European family developed a culture of its own with a provisional zenith in the European Renaissance of 500 years ago. They were capable of crossing the oceans of the world and laying hold of the wretches who had succeeded in travelling round the entire planet.

The discovery of two-legged, human-like beings on the other side of the Earth was a shocking experience for the Europeans because nothing of the kind was mentioned in the Bible. It was not immediately possible to recognise the kinship and thus it was unnecessary to show consideration towards these godless savages as fellow human beings. Many thousands had to lose their lives to direct violence and imported diseases, while their laboriously acquired items of value were plundered, destined to become part of the accumulated personal wealth of European princes. The advanced culture and the political system were smashed to atoms. Mind you, they had quite definitely developed in an insane direction.

Following the civilised clear-up, the old soldier buddies settled on their respective sides of the inaccessible Andes Mountain chain, and continue to speak the same language, even though they can definitely not pick up each other's accent. Now they sit calmly divided by the snow-capped peaks and snarl with resentment at each other. Who said the Swedes? Irish? French? Who said monkeys? Irrational hatred, moreover, does not exist in nature.

Chimpanzees are divided into two separate species: the little bonobo, which is also known as the dwarf chimpanzee, and the real chimpanzee which split from the same patriarch much later than the split from our own line. In genetic terms they are extremely close to each other. The chimpanzee is very aggressive towards foreign members of the same species, and will kill them if they come too close. Among the chimpanzees the victor will always triumph and rejoice, and the vanquished has to crawl – something with which we are highly familiar among ourselves. In the case of the bonobo things are reversed. If it defeats a rival the latter is offered reconciliation and friendship. It shows considerable empathy to suffering and grief, and also to other species. The curious factor is that the bonobo and the human being share some receptors in the brain which are believed to regulate and control the formation of pairs, social and territorial behaviour, while this type of receptor is not found in the chimpanzee. Why the human being exhibits traits from both species remains a mystery, but it is a fact that, in general, we do not care for strangers, we display distrust, or, at best, ignore them. We require ample time to create fellowship, tolerance and commonly held values, and empathy is equally well developed among us.

Handyman upright apes

The climate in Central Africa is hot and humid. Were nature to take control it would all end up as evergreen rainforest. Which doesn't leave many traces of people and animals for posterity, but, on the other hand, the apes have a wonderful home. Fortunately our forefathers' earthly remains turn up in new discoveries in countries such as Ethiopia, Kenya, Tanzania and South Africa, where the museums are increasingly groaning with skulls and implements – often reconstructed in exciting displays. It is however still an open question how far these reminiscences have brought us closer to a crucial image of the development of our line.

There is, as mentioned, a sort of consensus that the human line deviates from the line that later results in the chimpanzee and the bonobo, between four and five million years ago. Our line became the ape-like australopithecines or southern apes, of which we know eight or nine different species. They were small and hairy, almost like small chimpanzees, and thus not very different from apes. To a great extent one can discuss how far they represent a step in our direction. The very well-preserved and world-famous fossil 'Lucy' can be seen in a reconstructed version in the National Museum in Addis Ababa and represents a small ape-like dwarf which, according to palaeontologists, stands pretty far back in relation to the chimpanzees of today. The southern apes did not know they were in the process of becoming humans. They had adapted to that age's ecological conditions and could, theoretically, develop in any direction, just like the other apes.

About three to four million years ago our ancestors constituted a long series of apes which lived side by side in Central Africa. They were not very different from each other, but adapted to a number of different niches. The many different ones were, in time, reduced to a pathetic four species: gorilla, chimpanzee, bonobo and human. Along the way numerous relatives ended up as evolutionary aberrations. They perished for one reason or other, exactly like the ancestors of the present elephants.

The southern ape is an illustrative example of what we call evolutionary radiation, where a species has slowly developed into several species in various niches in the same way as elephants, bats and giraffes. The species of southern ape of which we have any knowledge, resemble each other and represent, according to established research, an interim stage between the ape line and the human being, even though, in fact, they continue to be mostly apes.

We generally divide them into two main groups: a group of relatively large and powerful species, which died out first, and a slenderer cousin that is believed to be the patriarch of our race, which diverged about two million years ago. All australopithecines died out at the latest one million years ago and thus must have lived side by side with the first member of the (the genus) *Homo* for at least a million years.

We do not know if the australopithecines were 'outcompeted' by the subsequent humans, or whether their brain could not keep up with the times and adapt itself to the capriciousness of the climate. They were actually witness to massive climatic changes throughout the million years when the climate underwent dramatic oscillations between cold and hot, wet and dry. A species in the wrong place at the wrong time did not have much of a chance. The *Homo* family, as is well known, managed to handle this upheaval.

The location of ancestors in an understandable, cohesive genealogical tree is an ongoing scientific discussion which does not always follow the urbane tone, nor is it always equally scientific. Personal prestige, fraud over fossils, national feelings as well as political and racist attitudes are all everyday fare, but let us confine ourselves to what we know with relative certainty.

Discoveries in East and South Africa indicate that the first ancestors migrated back and forth between these two regions through the East African Rift Valley, where some developed into independent enclaves, while others continued their migration and may have intermixed many times along the way. Informed guesswork, but far from improbable nevertheless. The known finds hardly cover their distribution, but they are merely an expression that the climate and the geological conditions at the find spots were of a preservative character.

The first southern apes were savannah apes, which left their colleagues in the rainforest, preferring the relatively open regions around freshwater lakes and rivers in the Rift Valley. What interests us most is, however, the leap from ape-like ancestors of the *Homo* genus, where the extended front limbs shrink and the brain grows. This splitting off happened from two to two and a half million years ago, after which the two lines *Homo* and *Australopithecus*, as previously noted, lived side by side for over a million years. Why did the *Homo* genus turn up? And how did it differ from the southern ape from which it is assumed to descend? Theoretically it could just as easily have descended from another species of ape, just as some of the southern apes could just as easily have diverged from the ape line – an extremely controversial and probably completely incorrect assumption, which is probably not so interesting either, unless one is a researcher. What continues to be the interesting point is: what created the *Homo* genus?

The discovery in Ethiopia of primitive axes and bones with marks of implements indicates that the use of the latter was introduced as early as two and a half million years ago and thus immediately preceding the development of the *Homo* genus. Animal bones with traces of stone tools indicate that the toolmakers ate meat and crushed bones to get the marrow. Fish bones are found in archaeological layers which are contemporary with the first prehistoric humans, which suggests that, early on, meat was supplemented with fish, and that these apes were somewhat more sophisticated than hitherto realised. This has led to a considerable number of researchers pointing to the protein-rich diet as the cause of the initial development of the brain. We know that our brains today function best with access to substances which are predominantly found in fish, so it is a logical deduction, but not quite in the spirit of evolutionary biology.

The average height of Europeans is said to have increased in the last 100 years as a result of better food, but it was actually the miserable food of earlier times which prevented individuals from deriving the full benefit of their respective genetic potentials. There can be hardly anyone who will credit fish-eating mammoths with being responsible for the large brains of today's elephants.

The complex and extremely energy-demanding brain was developed because it was an evolutionary advantage in this particular place at this particular point in time. What precisely the best brains of the time were used for we can only guess, but they equip the owner with the best 'fitness'. Better brains have led to better food and greater possibilities for further development. Thus the story of the chicken and the egg repeats itself. The brain expands from around

450 grams to 650 grams, barely half the size of the brain of the modern human being, but even so an increase of almost 50 per cent.

The intellectual and social functions of the brain both seem to be linked to its size in relation to body weight and to its inbuilt complexity. The larger and more complex the brain, the more advanced the life strategy with social functions, a high degree of learning and response to environmental factors.

The long history of birds has left species with brain structures which permit complex behavioural patterns despite the brain's relatively small volume. Here, evolution has solved flight's problem of weight with complexity. Better use is made of space. This can be observed, for example, in the crow family, which is capable of building further on experience already acquired. Learning builds further on learning. We do not know *how* this happens but we can state that it *does*.

The modern human being has a large brain in relation to its body and nature's average. This is a characteristic we share with the great apes, elephants, the great whales and the dolphins. Members of the big brains club are characterised by complicated family patterns, advanced communication, experience-based life strategies and the building-up of local cultures, where behaviour and communication vary from population to population. Chimpanzees which live a thousand kilometres apart from each other may have totally different cultures, different eating habits, diet and types of hunting.

It has proved that we large-brained species also share some physiological features in the brain, namely some specific brain cells (neurons) which are referred to by the abbreviation VEN and which come into existence in the thirty-fifth week of pregnancy. The cells continue their development until the child or juvenile is four years old when they have found their final place in three different locations in the brain. Since, as far as we know, they are only found in the aforementioned species and some close relations, such as the manatees (where they are, however, only found to a lesser extent) they are thought to play a significant role in connection with the aforementioned shared characteristics. We are all knowledge-strategists and can exploit the harvested experience for the optimising of the current ecological conditions.

Now it is extremely tempting to draw far-reaching conclusions, but a quick look back over the preceding pages can soon establish that factors such as the plastic structure of the brain, receptors and regulator proteins probably play just as great a role as the neurons, and putting it all together we are back with a picture of massive complexity which breaks with the old dogma that a gene is the same as a characteristic.

We have an idea that the first human being, *Homo habilis*, was the fruit of an *Australopithecus* – either *africanus* or *garhi* – and that it subsequently developed into the upright human *Homo erectus*, from which both the Neanderthal and the modern human being derive, even though we do not know the actual details. There also seems to be a clear line of development from the primitive hominids to the modern, civilised human being. A line along which the brain grows larger and larger; where technology is developed from primitive implements of stone to computers and space rockets, and along which the human being's ability to communicate verbally developed to a level which might be labelled sublime.

We see it ourselves as a cumulative development, in which we – present-day human beings – represent the top of the pops, and we expect that evolution will bring the human being even farther in the same direction. We become even more capable, more intelligent, so that in the future we will be able to conquer the universe. This notion is based on our own post-rationalisation, or rear-view mirror look at the development. We forget that mankind's progress is relative and that it only exists inside the human being's own consciousness. It is thus we ourselves who have defined the human being's development as a progress. In nature matters do not operate with such concepts as progress, only with adaptation. If I may be permitted to repeat myself, a gorilla is just as civilised as it can be. It is well adapted to its civilisation, and has the possibility of living a fantastic life in the jungle as long as it is not threatened by its so-called civilised cousins.

On the trail of the ape-woman's gorilla

The journey through the rainforest is longer and harder than expected. We are very sweaty, and the water has run out. We are still on the trail of the gorillas, but the group we have been looking for has nipped over the border into President Mobuto's Zaire. Crossing that line means trouble in more than one sense. The president, with the characteristic leopard-skin hat and Ray-Ban sunglasses, has no control over the frontier either, but he's a dab-hand at arranging world-championships in heavyweight boxing.

Our pathfinder has fortunately discovered the trail of the second group which is apparently sweeping the thicket in ever-wider spirals. We follow the trail for hours, but even though, at any point in time, we have hardly been more than 500 metres away from the group, there is nothing we can do except follow the trail of excrement, torn-down branches and nests. I use a much-needed break to make inquiries about the 'gorilla woman' Dian Fossey, to whom I had originally planned to pay a visit. We had been granted an audience at her research centre, though there could be no guarantee.

I remember clearly the paralysing shock which swung backward and forward between my toes and the roots of my hair on the Danish TV news's laconic item in the days between the preceding Christmas and New Year: "The well-known gorilla researcher Dian Fossey has been found murdered in her

Above: Well camouflaged.
Below: Bateleur eagle.

Above and below: Lions mating.
Right: Cycad at Entabeni. *Encephalartos eugéne-maraisii.*

An Umbrella acacia showing the silhouette of a Marabou stork during a grand African sunset.

Above: A modern-day man from South Africa.

Right: A modern-day man from Europe.

Right: Lucy, an *Australopithecus afarensis* who lived in East Africa three to four million years ago.

Below: A modern-day boy from East Africa

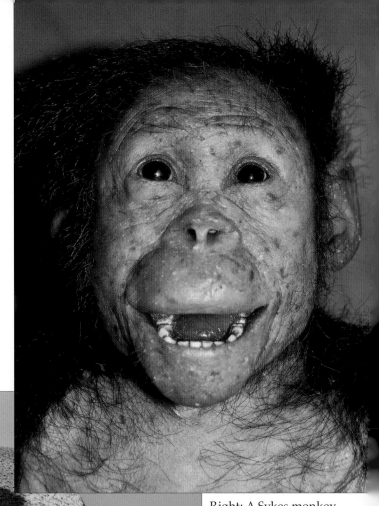

Right: A Sykes monkey.

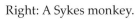

Above: A modern-day priest from Lalibela, Ethiopia.
Below: Swahili fishermen off the East Coast of Africa.

Black-faced vervet monkey carrying her young.

research centre at Visoke in Rwanda." The information came as I was sitting with my IBM golf-ball machine writing the itinerary for 'Gorilla Tracking' in Rwanda.

In the cloud forest the myths surrounding Dian live on in the best of health. There are good and inferior stories which are accompanied by a mass of theories about her death. An innocent student assistant was later convicted of her murder in absentia, but there is not too much to this story, neither have I met anybody who believed it.

Following a few years' study, Dian came closer to gorillas than any other human being. Her predecessor in the research post, Georg Staller, began his gorilla studies in 1959, but he never plumbed the same depths as Dian Fossey, who was prepared to sacrifice everything and set aside all personal considerations. This extraordinary American girl made the gorillas her immediate family and defended them implacably against every imaginable danger, which they themselves, as a rule, did not understand. Dian's notion left no room for the placing of the human being and the gorilla in different classes; on the contrary, the gorillas were better people, the company of whom was preferable to that of most others. This notion brought her into lasting opposition to fellow researchers and into direct war with all poachers and the capital interests behind them. She regarded the forests as her private domain and pursued everything and everybody, including poor forest people who set up snares to catch duiker antelope. This strategy did not make many friends, but made it easier for the corrupt politicians and conscienceless moneymen who sell gorillas to zoos, where the young waste away and die of grief in the space of a few months. Gorillas are not suited to life away from their families.

Dian Fossey divided her time between research and the hunting down of poachers, and anyone who wanted to study gorillas and write scholarly papers at the Visoke research station was told to plan their time on the same lines. Meanwhile, the civil servants in distant Kigali sat and fulminated over this difficult woman who was destroying a reliable source of income. Commotion was something she raised a deal of, and it was widely heard in Copenhagen, Paris and Washington.

Everyone in our modest circle knew Dian Fossey. One had been directly appointed under her at Visoke and two park wardens worked almost daily with her in the hunt for poachers. Nobody had anything but good things to say about her, which may possibly be due to the fact that gorilla tourism has now secured everyone a regular income and generous gratuities (tips), but whenever I go into her close relationship with the apes – I use the word family – then the laughter starts. Of course it's amusing to regard apes as closely related to the human being. Just like a type of cousin. Can you eat apes? I inquire. Of course you can't, but the denial is of a piece with the fact that it's forbidden: neither would gorilla tourists be inclined to reward that kind of utterance.

They are, however, aware that the forest's uncivilised pygmies eat apes, if they can catch them. Clear documentation, isn't it, that these people's character is truly of 'lower standing', and quite definitely closer to the gorillas than real people in brick houses? You live and learn.

The gorillas of the forest are often called mountain gorillas and are considered to be a special sub-species, even though it is unclear which is the 'sub' and which the 'upper' species. All gorillas are the same genetic species, but have local adaptations exactly the way Europeans, Eskimos and Chinese are the same species as our pathfinders, and, if fair is fair, the species *Homo sapiens* also includes the ape-eating pygmies of the forest. In recent years we have regarded the gorilla as a shade more primitive than the chimpanzee, which works out as being a step farther away from the human race, but one could also cite the point of view that the gorilla's line of development is older than that of the human being and that therefore it has had longer to fine-tune its development in harmony with the environment.

Mountain gorillas live in dense, inaccessible cloud forest at an altitude of

3,000 metres. Here there are no enemies or significant fluctuations in climate which affect the supply of food to an appreciable degree. Gorillas breed slowly and the population seems to have adapted to the supply of food, which is made up of the leaves, bark, twigs, roots, flowers and fruit of about 150 species of trees, bushes and shrubs. In some cases it eats anything, and in other cases it is more selective. It can be content with foraging in the hours of daylight and yet have plenty of time for a nap once in a while and social contacts. It has, in other words, plenty of quality time, something which the modern human being seems to lack.

The gorilla knows its diet in detail. It knows when there are fresh bamboo shoots in the lower-lying bamboo forests, and where and when there are particularly sweet fruits or especially juicy leaves, for which its is worth going that extra little distance. There is, in other words a great deal of knowledge involved with being a gorilla, knowledge which they acquire from an early age in the daily gorilla school. When alpha males run into each other unintentionally they tend to avoid life-threatening combat but manage the hostilities in splendid masculine style.

If the concept of absolute adaptation has any meaning it seems undeniable that the gorilla has found the answer. There is no pressure on the species from predators, no shortage of food nor overpopulation. It is completely and utterly adapted to its habitat and can, as previously mentioned, spend a good portion of its existence in relaxation, good fellowship and social gatherings. The gorilla has understood how to arrange matters much more comfortably than the eternally warring Hutu and Tutsi peoples who surround its country.

Before Darwin it was an indisputable fact that animals have no soul as Descartes formulated it. The distinction between animals and humans is completely unequivocal, as spirit comes from God and without it one is not a human being, but put on Earth to serve him/her. Darwin brought order into the system. If we have developed from the same origin, where and when does the spirit or the soul come into the equation? Anyone who spends time studying hominids and a whole lot of other animals besides seems to end up with the same result: they are unable to set a boundary between people and animals. They are not in a position to point out the crucial factors, which establish this division once and for all. Of course there are differences. Apes don't write books or manufacture flat-screens, but does this justify talk of difference? We are, though, in agreement that human beings possess certain peculiarly human characteristics. In that sense are we human and not animals. We still lack that ability to define the peculiarly human qualities which make a difference.

The London professor, Felipe Fernández-Armesto, writes in his little book, *So You Think You're Human?* that, objectively, chimpanzees and humans are so alike that anthropologists from Mars would classify them together, and would thus be completely in accord with the scientists who call us 'the naked ape' or 'the third chimpanzee'. In the book he also has his colleague from Princeton, Peter Singer, argue scientifically and passionately that the net enclosing the human race *Homo* should be cast wider so as to also include chimpanzees. Why not gorillas, baboons, elephants and killer whales? In what way are chimpanzees specially qualified? Would it not be more interesting to discuss what one should use this demarcation for? Pigeon-holing under the category *Homo* brings with it automatic granting of rights, human rights, enshrined in law and entered into the statute book, which, however, neither in nor outside this category, seems to make a tremendous difference.

Since Darwin we have been seriously plagued by this nagging doubt and have fought a brave battle to safeguard the distinction between animal and human being, but the more we delve into the subject, the more differences are erased. Every owner of a pet knows their own animal's qualities: smart, intelligent, clever, affectionate, helpful, lazy or stupid. For precisely this reason it seems important for us to find the peculiar 'human' factor x that separates us from the animals: the argument that we can treat animals as inferior, as food and slaves. It is somewhat easier to eat a good red steak, if the ox (from which it came) was unaware of its fate, but in terms of cognition we still have problems with ourselves, our origin and what we regard as lower and higher species.

We define our ancestors as human-like beings, which developed between two and two and a half million years ago. In reality these first humans were not that different from southern apes but we believe we can trace a certain similarity to the present-day human being. We have discovered a number of fossils of these prehistoric humans, which we call *Homo* with appellations such as *africanus*, *rudolfensis*, *habilis*, *ergaster* and *erectus*, of which the last-named is believed to be our own ancestor. *Homo erectus*, the upright man, is presumably the closest ancestor to both us – *H. sapiens* – and our close relative *H. neanderthalensis*, with whom we have mixed blood.

There is still dispute over the actual sequence and dating, just as one would expect to find a number of links hidden below the surface. Absolute cast-iron proof of descent is something we do not have, but the overall idea of a step-by-step development makes good sense. A quite solid scientific theory which makes the finds hang together, so we'll snap it up for now.

Homo erectus managed to leave Africa a million years ago, bringing with him fire, weapon technology and probably good hunting skills. This was an extremely capable race or species which understood how to look after themselves across most of the planet and under the worst climatic conditions imaginable. Some people believe that they were very close to us in intellectual understanding, but this is something about which there is difference of opinion, if not violent disagreement. *Homo erectus* probably wandered out of Africa in several, possibly many stages and therefore it is naturally difficult to establish whether these people were just as capable a million years ago as they were half a million years ago, and when the various developmental features occurred. There is a basis for theories, but this doesn't change the fact that *Homo erectus* was somewhat quicker on the uptake than us. Long before present-day human beings our upright ancestors managed to spread throughout the entire Middle East, the Mediterranean, Britain, the East, the Far East and Siberia and keep the species alive, until we turned up on the stage, or close to it. *Erectus* lived simultaneously with the Neanderthals, who lived at the same time as we did. It is not improbable that all three species lived side by side.

The traces of our ancestors and early human beings are unmistakable. We keep digging up new fossils and tools from the bogs and swamps of the entire world, and together they increase our knowledge. The first Neanderthal finds turned up in the first half of the nineteenth century, and then the first *erectus* was found in Indonesia by Dubois in 1891. Various species from various eras and discovered in various places led to the persistent theory that the human species had developed from various lines, in various places on Earth. Kinship between white, black, yellow and any other possibilities, was, in other words, pretty insignificant and extremely remote: a comfortable and acceptable conclusion in most European, or rather white, societies.

In the latter half of the twentieth century, however, the 'out of Africa' theory began to exert pressure, to the considerable misgivings of fundamentalist religious circles, which for a long while were seconded by serious researchers. The theory stated that all the human beings on the planet developed from the same line in Africa. All people outside Africa come from a small group which once wandered out of Africa. Everybody, in other words, is each other's cousin.

Today the evidence is so overwhelming that one would have to be ignorant or worse to maintain that each of the peoples of today sprang into existence by itself. That would fly in the face of the theory of evolution, genetics, palaeontology and every imaginable system of logic. Genetic investigation can, with exceptionally high precision, demonstrate that all people outside Africa descend from the same, small, wandering group which took a firm grip on the 'new world' around 60,000 years ago. In advance of us, early ancestors such as *Homo erectus* and Neanderthals also wandered out, but they are all extinct. All Europeans, Chinese, Aborigines, Native Americans and Inuits are very closely related – a little more closely related to each other than to the African relatives we left behind. Ultimately we all descend from a relatively modest gene pool which had its origin 120,000 to 200,000 years ago in Central Africa. Genetically we are still more closely related to each other than to the apes, inasmuch as that makes any difference.

The vanished tribe

We have heard about poachers, corrupt politicians and police and other people of lamentable character – exactly the kind gorillas don't care for – but we don't see them, or else they've also forgotten the remarkable sign on their forehead. People are kind and when, along the way, we wave to a working man sweating the proverbial buckets, he waves back with a smile. My extremely broken French is received with a smile and sometimes with a good laugh. All in all there is not much to complain about, as far as human friendliness is concerned.

Lurking beneath Central Africa's peaceful exterior however is something all but idyllic. At regular intervals leading ethnic groups are seized by an irresistible urge to exterminate each other. Principally without any apparent reason. Conflict in the local sense is called genocide in European. A local holocaust with the sole purpose of eradicating another ethnic group, simply because it is there. According to the international press coverage, the battles are between the numerically superior Bantu group, the Hutus, who make up 85 per cent of the population and the tall, slender Tutsi or Watutsi people, who, through the last fifty years have been butchered in their hundreds of thousands. The latter group are, however, fierce soldiers who tend to return with relentless force once they have assembled the troops.

The Hutus are apparently so terrified of the strength of the Tutsi that a takeover is instantly exploited to be rid of the problem, while the Tutsis are well aware of the danger and attempt to hang on to power with military force. Once in a while peace and tranquillity aspirate. Then one is entitled to talk in terms of an idyll.

The fighting ethnic groups speak the same language, are solidly intermarried with each other and cultural differences are not readily apparent. For years they live closely together, apparently without daily hostilities, but then suddenly, one day, all hell breaks loose. Hutu groups rage through the country, annihilating everything in their path. Tutsi women, children, babies and sympathisers, those who have intermarried or friends. Everything must go. A deep-felt pathological hatred flares up like a bolt from the blue. How can it be explained? Rwandese politicians and researchers trumpet inflated theories that the seeds of hatred were sown by the colonial powers on the ancient Roman principle of 'divide and rule'. We have played them off against each other in order to plunder the land. Naturally this contains a grain of truth but it doesn't get you around the fact that every single Rwandese identifies him or herself as belonging to either Tutsi or Hutu (or the pygmy people Twa). Perhaps it is not exactly the basis for genocide, but the powerful sense of affiliation is a fact when all hell breaks loose.

Visitors see nothing of the hatred that smoulders beneath the surface. We travel around safely and are met, as I said, with friendship and abundance and can concentrate on judging the accursed poachers who murder the mountain gorillas. Neither do we know that we are standing centre stage where one of Africa's most appalling acts of genocide will take place ten years later, between our visit and this account. We know about the past but, as with most of the African peoples, history blurs into the uncertain when one goes 100 years back. Where do the various physical traits come from? How have groups migrated around the continent? Perhaps they have travelled outside Africa in the process. What are the grounds for disagreement? We may have to wait a number of years before the genetic research gets the better of the complex history of migration, but here is a shadow of one of humanity's oldest, classic conflicts: that between nomads and arable farmers. Two directly opposed strategies, each of which is entirely suited to its own environment and tremendously conflict-filled in their common milieu. A conflict which goes right back to the Pentateuch, to the clash between the kin of Cain and Abel.

Within our little group there is no doubt whatsoever. There are Hutu and there are Tutsi, which, however, does not seem to make any great difference. If we ask people on the roads the answers are fairly unequivocal. The difference does not appear too striking, but we do see Tutsi with a physical build which prompts thoughts of the peoples who migrated into East Africa from Ethiopia and Sudan as Nilo-Hamitic pastoralists (cattle nomads) The Tutsi call themselves cattle people while the Hutu style themselves farmers. In practice, however, this distinction broke down long ago but the words remain the same. A Tutsi is a Tutsi and a Hutu is a Hutu. So the anthropologists can think what they like.

The Tutsi were originally cattle people and cattle nomads and related to the Masai and Samburu in Kenya and Tanzania. Nomad cultures have developed in the world's agriculturally marginal areas: deserts, semi-deserts, sahel and savannahs, where there was no alternative to arranging one's way of life to be compatible with nature and migrate according to the periods of rain to obtain sufficient food: a technique which Africa's wildlife learned long before the arrival of human beings. Genetic investigations have established that we tamed many of the domestic animals around 12,000 years ago, roughly the same time that we learned to cultivate the soil, raising and developing crops. This was about 40–50,000 years after the modern human beings' first successful migration out of Africa. Africa's cattle nomads do not resemble her other peoples which are primarily made up of Bantu, who have a shorter, stockier physique, belong to another language stem and are arable farmers. The cattle nomads have Indo-European features and a fundamentally different culture, a different religion and a different language. The spread of the cattle nomads drives a number of wedges into the Bantuland of Central Africa, and we know for certain that they migrated in from the North, fighting for land and position along the way. My guess is that all the cattle nomads of East Africa come from a people who moved out of Africa and migrated around in the Middle East, where they acquired Indo-European facial features and build, learned cattle farming together with the peoples who later migrated northwards and lost skin colour. Six or seven thousand years ago they returned to Africa with their cattle and gradually spread out along the lush rivers of the Sahara to the south and east, as the Sahara dried up. Some ended up as Fulani in West Africa while others ended their journey eastwards. A study of the development of climatic conditions and ancient water courses makes it possible to trace the migrations in detail, supplemented by discoveries of stone tools and cave paintings.

The most key argument in favour of this theory is the different physical features of the nomads and the Bantu, and particularly the milk tolerance of the adult nomads, which is due to a late mutation. All mammals' milk contains the sugar lactose which is broken down in the body into easily digestible sugars. This breaking down occurs with the aid of an enzyme (protein) called, after its function, lactase. When breast-feeding stops, mammals have no further need for the ability to break down lactose, and production of lactase ceases. Evolution does not permit us to waste energy on something for which we have no need. Therefore the organism simply shuts down the gene which produces lactase immediately on the cessation of breast feeding. In 75 per cent of the world's population lactase production is cut by 90 per cent or more. The milk, therefore, in an undigested state, enters the large intestine where the bacteria take care of the metabolising, in the course of which there is the development of large amounts of gas.

Cattle nomads have a mutation in chromosome number two which prevents the shutting down of lactase production. The mutation presumably occurred with the first cattle users more than 10,000 years ago and has been contributory to the development of cattle farming and cattle nomadism – and/or animal husbandry has favoured this mutation. Cattle farming, soil cultivation, hunting and gathering clearly supplemented each other in the dawn of civilisation, but the functions have become increasingly specialised, and at some point families with the favourable milk-tolerant mutation split off and discovered their own path. We know that the Tutsi population carries this precise mutation. Some believe, however, that the mutation or mutations emerged independently of each other, which may be the case since different mutations may have the same, or very similar, effects. In this case, however, it certainly seems to be a case of the same mutation.

Provided this assumption is correct, it should be easy to establish because Europeans are so much more closely related to Africa's cattle nomads than to the remainder of the continent's population, and cattle nomads are much more closely related to each other than to the Bantu population of Africa.

Gradually, as the nomads have spread out towards better and more fertile areas, they have clashed with the arable-farming Bantus, and the conflict has become a reality. Through history the nomads have prevailed, apparently because they could replace the hard toil of field work with training for war. Most of the nomadic people in Africa, such as the Tuareg, Masai and Samburu are actually characterised by a social organisation with a military caste.

The Masai were, and still are, legendary warriors and their reputation is so terrifying that they have seldom had to actually use their skills. For several hundred years a numerically small band controlled the best agricultural land throughout the whole of East Africa, without the arableists standing a chance of penetrating into the interior. This was first achieved with the help of the Europeans.

It could be interesting to shed some light on how far the warlike nomad people's mentality has genetic markers. Their development is a lengthy process, which has set its clearly genetic mark on the external features, but might there also be special internal features?

As mentioned previously, the Tutsis are usually defined as cattle-people, and the Hutu as arable farmers. The two groups ran afoul of each other in the fertile regions of Central Africa several hundred years ago, and this clearly wasn't without bloodshed. It was this fertility which formed the ground for a common system in which both parties co-existed with the numerically small, warlike Tutsi group as an African aristocracy, or order of knighthood, which ensured the hair-fine balance through physical superiority. The balance was only upset with the arrival of the Europeans and a democratising process whereby the numerical superiority of the Hutus ensured their power.

If one reads the reports from Burton's, Grant's and Speke's journeys in the search for the source of the Nile, the kingdoms in Central Africa are presented as agricultural societies, governed by an isolated aristocracy of people of contrasting appearance: tall and slender as opposed to the short, powerfully built tillers of the soil.

When all is said and done, it doesn't take much imagination to figure out where the real problem lies hidden.

No flawless monkeys

Finally we enter the gorilla family's sitting room. We sit down, look at the ground and show humility. The gorilla family is quite unconcerned. Dian Fossey and her team have really blazed a trail for the rest of us. We sit still while the gorillas themselves make incursions within the recommended critical distance. A full-grown female lies in the thicket less than a metre from me. I cannot touch her, but I can easily smell her and hear munching noises, gentle snorting and stomach rumbles. Then a small infant pops up and crawls over her stomach. It takes a nip or two among the leaves and we can therefore conclude that it must be about four to five months old, even though it doesn't look its age.

Gorillas carry their babies two weeks less than a woman, and its juvenile is a couple of kilos less at birth, but somewhat faster onto its legs than the defenceless human child. Perhaps I am biased but, to a high degree, I see a family and a family idyll. The scene overflows with cosiness and well-being. The old Silverback – the undisputed head of the family – remains at a slight distance, but scowls at us once in a while. I interpret the look as one of slight irritation. It's difficult to forebear but it is definitely not a scientific observation. The clicking of cameras is visibly disturbing, and I relive the eternal conflict between enjoying the moment and perpetuating it. It is difficult to get good pictures in the dense thicket. Silverback will forever escape, but, on the other hand the young one poses on the stomach of its mother. All of them are very preoccupied with feeding, which, however, comes to an abrupt halt, even though the food seems abundant. Perhaps they're tired of us.

Long ago Dian Fossey and other researchers established that gorillas communicate intensely with each other. Dian could easily influence their behaviour by imitating sounds and attitudes. Right this moment we communicate by displaying humility. It is quite clear that the signal is understood. It is also clear that there is a lot of interaction between the members of the group, but it will take many years behavioural study under difficult conditions to interpret movements, expression and sounds, and so we must live with the cultural limitations which suffuse our view of the world. Might it be that the apes have a god?

With the small and more primitive savannah monkeys which could formerly be bought in pet shops like guenons, we can recognise at least ten different sounds signifying danger. Every predator, such as leopard and Martial eagle has its own sound in ape language. The use of speech is even sufficiently advanced that they can use the sound of a leopard, for example, to frighten off attacking members of their own species. Following studies and video-recordings from San Diego Zoo, the researchers Waal and Pollick identified eighteen different facial and sound signals as well as thirty-one different hand and arm movements among chimpanzees and their close relatives, the bonobos. They could, in addition, state that this communication was far more complex than previously supposed, in that the same sign could constitute different messages, all depending on the situation in which they occurred.

There are numerous examinations of individuals among the anthropoid apes, which appear to reveal capabilities for complex thought processes. There is, however, considerable disagreement over the interpretation, which ranges from one extreme to the other, and all points in between: from the very human to the opposite, whatever that may be.

When a large female is lying in front of you and munching on a handful of leaves, while cosseting an infant who is crawling around on her belly, and the gigantic father stands in the background acting as though nothing is happening, while casting surreptitious glances in your direction, you invariably speculate about what is going on in their heads. Do they think? Do they talk? In any case they appear to have reached agreement on the relaxed attitude they have adopted towards us.

A miserable twenty-five minutes is all that the gorilla family grants us before disappearing in the green depths. I would gladly have followed them, something which brought roars of laughter from our pathfinder. This idiot thinks he can follow gorillas. Forget *all* about it. They want some peace and quiet, and we haven't an earthly chance of tagging along – you don't get to be a gorilla in twenty-five minutes. The trek down takes only half an hour so the gorillas, one might say, have led us up the garden path. We have walked around them for hours. Exhilarated over the experience and a mite irritable that the visit is over, you get greedy, no doubt about that. Any wiser?

It's impossible to observe gorillas for many minutes without getting the feeling that one is watching something familiar, that is, if one can otherwise turn one's attention away from the abundant hair. Right? What is it that separates them from us? Or, rather, what entitles us to set ourselves above them? If we possess peculiar characteristics and feelings, wouldn't that result precisely in us being capable of showing respect towards our fellow creatures?

All creatures have undertaken a common journey through evolution. At one point in time some of us became capable of developing our embryos inside our bodies and feeding the newborn with our own milk, which has made us tremendously more mobile and enabled us to occupy new niches around the planet. Much later a small group took possession of the tops of the trees and among these a small number have developed a flat snout and eyes located side by side. This brought about stereoscopic vision and the ability to hop around unhindered in a jungle landscape. These became gorillas, chimpanzees and human beings, while some of the ancestors remained shrews and lemurs.

According to Darwin, it's conceivable that some individuals were born with genetic faults and deformed faces, which have presented new opportunities to occupy new food niches, or thrive better in the existing ones. A development was initiated, but this was not without some cost, as a small snout meant there was less room for olfactory receptors, and the sense of smell was quite significantly weakened. A smaller snout also meant that the body's ability to regulate temperature through the throat was reduced, and the individual had to compensate for this through sweating. In this way this organism cut itself off from a life in dry regions. Regulation possibilities were further reduced by a palate, which, on the other hand, increases the capability of communication through speech. Thus our common development as, and into, anthropoid apes had its ups and downs, which has the common feature that precisely the characteristics which slightly increased one's ability to survive, were prioritised by natural selection and nearly always at the cost of other possibilities. The current environment alone determined the direction of development, and every ability or value of the characteristic only counts in relation to the surroundings. On the drawing-board of evolution there are no peculiarly human x-characteristics, and everything indicates that we alone are driven by the possibilities of evolution.

All human beings have different genes. The people who live outside Africa seem to be more closely related to each other than to the population of Africa, which indicates that the successful migration from Africa happened just once. Even though we are so close to each other in a genetic sense that we recognise ourselves as belonging to the same species, which is, therefore different from gorilla, chimpanzee and other animals, this does not prevent us from killing each other without much profound deliberation. No matter how closely related we are or how much we have shared throughout history, it doesn't seem to have made a lot of difference, does it? One is tempted to ask to what degree we have anything particularly desirable in our gene pool that our hairy relatives do not possess, and as to whether, when it comes down to it, the price was worth paying. There is always a price.

The third chimpanzee

Is the human being a third chimpanzee, which has become more civilised, and has acquired new human characteristics in a short time? This is sort of the content in the superscription and title of Jared Diamond's fascinating book, in which he advocated the exciting and still controversial theory that the human being underwent an intellectual leap 40–50,000 years ago. According to this theory, the modern human being was not fully developed when it turned up on the world stage 150–200,000 years ago, but had to wait a further 100,000 to 150,000 years for the great leap which put it in a position to develop the world into how we see it today.

A good 40,000 years ago tools changed. They became far more sophisticated, and their individual purpose quite clear. Our ancestors – in Europe Cro-Magnon Man – sewed clothes, produced fishing nets, harpoons, bows and arrows. They adorned themselves with beads, drew beautiful pictures in the caves of Southern Europe, introduced burial rituals and began to resemble the human of today almost to the point of interchangeability. Several books on the subject have already been written on the subject, and one of the most famous proselytisers is Noam Chomsky, who, with a background in the human being's current language skills, concludes that language is a result of a small genetic change, and a lightning development of 40–50,000 years duration. The theory has met with masses of resistance from the scientific world, where it is maintained that the development of language skills and intellectual characteristics is the result of a lengthy development from the childhood of the modern human being where we even build further on the skills of prehistoric human beings. We have in consequence two diametrically opposite theories based on a short and a long evolutionary adaptation respectively. The latter has a certain inner logic because we naturally build the entire time on something previous, a fact which is also supported by the traditional Darwinian mode of thinking about the laborious evolutionary development in small bites.

One cannot however turn aside from the fact that archaeological discoveries document a rapid technological development. Why, in a short time, did we suddenly become cleverer? Why did we use the same stumpy stone hammer for all sorts of purposes for a million years, when in the course of just 10,000 years we can learn to grind it into a super-sharp weapon, fix it onto a wooden haft and produce new variations on long shafts, small knives and numerous tools for quite specific purposes? There is a point which, at best, may mean that both parties are correct.

In the aquarium outside Copenhagen they had a smart octopus which was capable of unscrewing the lid of a jam jar to get a fish which had been put inside it. My friendly introductory inquiry,as to whether it was an intelligent octopus was rejected by the professional guide in the most definite tone. The board of the institution has actually directed him to think that the animal is not intelligent. This characteristic is exclusive to human beings. I don't know what one is supposed to call it in an octopus, but it does beg the question.

The human being is intelligent, that we ourselves have established, and we have supposedly become more intelligent along the way. Is it perhaps intelligence which has developed and given us language? The problem is that we have formidable problems with the concept of intelligence. On the one hand it is a tremendously praiseworthy and prestige-bestowing characteristic, of which, apparently we cannot get enough. On the other hand large areas of science are reluctant to handle the concept, because it has been used in extremely controversial and racist contexts.

First of all we measure something unknown with a test, and afterwards we attempt to find what it is we have measured, and call it the intelligence quotient. The test has been 100 years in the making since the psychologist Charles Spearman noted, in 1904, that there was a correlation between the school pupils' abilities. To put it simply, pupils who were really good at one thing were, as a rule, also good many other things: the characteristic he called intelligence. In time the whizz-kid researchers into intelligence refined and adjusted the test and, in the process, purged it of social, cultural and educational sources of mistakes. Today the test can be repeated time after time and still give the same result.

It begins to resemble science.

The interesting thing about intelligence is that it is in fact the only mental characteristic we can measure. We cannot measure the degree of empathy, love of one's neighbour (charity), motherly love and social attitude. Simply put, that's just about everything which, in our opinion, makes us human beings.

Today scientific research into intelligence deals with the very narrow characteristic which was described by Spearman and is called the G-factor. The G-factor is measured with the so-called IQ or intelligence quotient, which, roughly speaking, describes the ability and speed wherewith we are able to see through complicated abstract contexts. The G-factor is markedly inheritable, remains at the same level throughout one's life and is critical for how we accomplish many of life's tasks.

People with a high G-factor are, as a rule, good at both languages and mathematics, but can also score high in music, creativity and social awareness. Intelligence also says something about how good one is theoretically at accomplishing tasks and acquiring skills. The G-factor and IQ which we're talking about here are thus something quite concrete which has nothing at all to do with the age's 'woolly' concepts such as emotional intelligence and social intelligence, which, with all due respect, do not have the same empirical foundation.

Chinese and Japanese on average have a 5 per cent higher IQ than Europeans, whose IQ is far higher than the average among Africans and South Americans. These average figures obscure enormous differences. In practice the ten most gifted people may well turn out to live in Africa.

Regardless of where in the animal kingdom one finds intelligence, the interesting thing about the characteristic is that it is inheritable and therefore affected by evolution. Average differences between population groups, such as black and white, men and women, are not especially interesting, but it is, on the other hand, the causes of such differences and the other characteristics which fight for prioritising along the road of evolution. If evolution has used a great deal of energy to develop intelligence, what has been the price? If a little energy has been saved on the development of intelligence among other, possibly less intelligent species, what has been the result?

This situation can tell us a great deal about the human being's cultural development. Why have Chinese and Japanese higher intelligence than Europeans? One explanation may be that their languages are far more complicated and difficult to learn. Despite the difference in intelligence, it still takes a Chinese school pupil longer to reach the same linguistic level as a European pupil. He must first learn up to 2,500 complex written characters, where a Dane can be content with an alphabet of twenty-nine and an English pupil with only twenty-six letters.

This is of course a guess, but one can easily imagine that evolution has favoured the linguistically best-developed. Anything else would look weird. The question is, then, what have the Chinese paid for their extra portion of IQ?

Regional differences in intelligence are due naturally to the fact that evolution has prioritised various characteristics in relation to current needs. But wait a minute. If this supposition is correct it means that intelligence is a pretty mobile characteristic which can be altered or influenced in a relatively few generations. A thought which has some difficulty in gaining a foothold. I shall return to this presently.

The usability of intelligence can be interpreted in many ways. Tough tribes who live in harsh, dry areas need experience and memory. One learns everything about the surroundings: what is edible, when, where and how, how one survives in marginal regions when it's necessary. The message is: learn, remember and grow old, so that others can benefit from that experience. In this context, some will have recourse to conservative life strategies and have no need for a creative, developing brain, while others migrate out of Africa in order to try something new. These needs are extremely varied, but what they have in common is that various solutions exist to the problems, and that no one solution is more correct than another. It is not the individual who lives longest who has exerted the crucial influence.

The point is, therefore, that intelligence can show itself to be an outstanding characteristic, which can jump-start an innovative development, such as the industrial revolution that happened in Europe, but there are also other possibilities of successful life strategies. Measured against nature's yardstick, success is synonymous with survival and reproduction, irrespective of whether this requires goat's milk or Airbus.

Naturally science has searched high and low for the inheritable intelligence factor which one is convinced exists. Today, however, it is relatively clear that no such thing as a specific intelligence gene exists, but that people with high intelligence have many genetic variations, each of which apparently contributes a small amount to 'the' intelligence. Accordingly, a large number of genes contribute to increased intelligence. Or it's something else we don't know.

What is the significance of these minute genetic differences? One point six per cent turns chimpanzees into human beings, or vice versa, but a glance at dogs reveals that all races of dogs are one and the same species, with a genetic uniformity of 99.85 per cent. We began to tame the wolf some 15,000 years ago and about 4,000 years ago we took the next step and transformed the tame wolf to serve every imaginable purpose: thus we have hunting dogs, herding dogs, guard dogs, fighting dogs, tracker dogs, lap dogs, dachshunds and many others. In a few thousand years it has proved possible to change the appearance of the wolf or dog radically, in terms of size, weight, musculature, colour, length of hair, ears, muzzle and face. And it doesn't stop here, because we have actually also been able to influence the dog's temperament, learning skills or intelligence, social adaptation, and intuition. A modest 0.15 per cent can bring about a leap from a small good-natured dachshund, which lies wagging its tail in the Danish Prince Consort's lap, into an obstreperous, powerful pit bull terrier with a bite like a hyena and the temperament of a psychopath.

Recently an interesting experiment was performed in Russia at the University of Novosibirsk, where an attempt was made to produce a tame dog starting from a silver fox, the behaviour of which recalls that of a wolf or a wild dog. The assumption is that the taming of dogs and other tame animals occurred over a period when the least aggressive and 'human-friendly' individuals were selected for breeding. The experiment was carried out in the same way and resulted, in the space of just eight to ten generations, in a tame dog. The artificial selection brought about numerous hormonal changes, which were slowly stabilised in step with the domestication. The signal substances of the brain changed and reached a stable, steady level after the enumerated eight to ten generations, but beyond this numerous other alterations occurred in the regulating of the animal's genetic expression, which were reflected in the respective level of hormones in the blood.

We know that an evolutionary pressure in a specific direction can create relatively rapid genetic alterations on a physical level. This applies in the case of skin, eye and hair colour, but there is the increasing indication that the same can happen when it is a case of mental characteristics. If we need intelligence, empathy or aggression, it may be developed very, very fast, if the pressure is sufficient. The human being has DNA receptors in the brain, in common with the bonobo, the receptors which regulate the bonobo's social behaviour and apparently create a non-aggressive, empathic and peaceful species. Theoretically we ought to behave in a similar manner but we have evidently had need of a partial deactivation of the system, since we can even behave with extreme aggression. The question is if this applies to all human beings, the species in general, or whether there are in reality several distinct differences such as can be observed with intelligence.

Have the warlike nomadic tribes built up a genetic disposition as warriors? East Africa's cattle nomads appear intelligent, peaceful and only slightly aggressive, but we know from history that there's a short fuse when they are provoked and that they can go into character as utterly incomparable warriors who are so many classes above all the others that it is staggering.

On closer inspection it is clear that the human being has perhaps various marked characteristics concealed in the modest 0.3 per cent difference, which occurs between our genomes, and which may conceal 100,000 different mutations. Perhaps the difference covers an important regulator. If so, that would mean that we are capable of influencing our own mentality in the course of a few generations. Who knows – perhaps we are developing into an empathic, peaceful and tolerant generation?

Not before time!

Crowned crane.

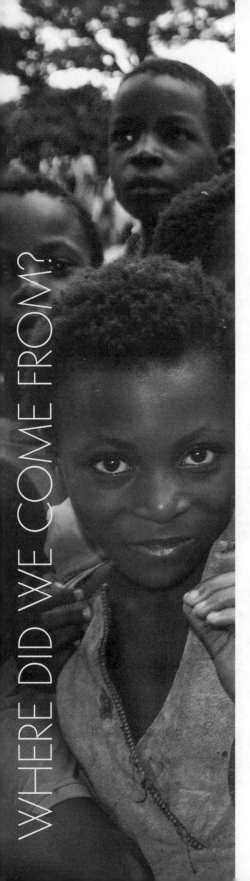

Humanism is the Sister of Science which ensures that Scientific Progress is Progress indeed.

Bagamoyo, Tanzania, 2009

Coconut palms stretch as far as the eye can see along the gentle curve of Bagamoyo's coastline where the white sand beaches separate the Indian Ocean from the fertile land. Waves of the incoming tide wash over the protective coral reefs, while outrigger canoes and small dhows pitch in the rising waters. We are standing in a picture from a tourist brochure even if the exotic set is disfigured by washed-up plastic bottles, tin cans and coloured nylon rope.

The older men of the town have found a place under shady trees, where they remain seated for most of the day. Children and adolescents drift around at random in the bleak surroundings, where there is little of any consequence to do. Nobody makes any more wooden elephants, Masai warriors with spear or outrigger canoes than the market demands, and the market is right down. And that is the case with most things. The sea is rich and some people fish, but the amount of surplus fish is already overly large. There's no demand for more. The earth in the hinterland is full of coconut palms, alternating with mango, papaya and other crops, which can go straight in your mouth without expending much effort. The gruelling toil in the fields is no work for a free man. You don't get to be free through slave labour, do you? The past clings stickily to the present. Slavery and the old norms adhere.

Those slaves who had bought their

242

freedom established their own coconut plantations, and did not hesitate to exploit other slaves. That was just something people did. Although the conditions for the slaves gradually approached those of miserable wage-earners, servitude was discontinued around the turn of the century and the First World War under pressure from the outside world. New solutions had to be found and work in the fields became the preserve of the women.

Human beings inhabited Bagamoyo for a long time before the name was devised, and long before we became Arabs, Chinese, Europeans and slaves. We – that is modern, smart people and our ancestors before us – have clearly wandered around East Africa and lived on this stretch of coastline since time immemorial. Here there is fresh- and saltwater fish, fertile soil, fruit and vegetables and animals you can eat, and so it has been for a very, very long time, even though tidal waves, drought and volcanoes in the hinterland have sent the inhabitants out on temporary migrations. One of these may have led people out of Africa.

For over a thousand years traders have come in search of ivory, termite-proof mangrove sticks, limestone and exotic hides and spices, though slaves probably constituted the town's, even the whole of Africa's, most important trade commodity for an even longer period of time. Millions of people were transported against their will away from these stretches of coastline and many died during the hunt for slaves.

The final full stop in the grotesque slave-history of Bagamoyo was inserted in 1977 when the last slave died: an old woman, born in slavery, the daughter of a plantation slave bound to the soil. The German colonial masters who took over the town at the end of the nineteenth century, silently accepted slavery until they themselves were thrown out after the First World War. What else could one do with these people? The slave-owners were themselves freed slaves who gratefully delivered the commodities and taxes to the master race.

Bagamoyo today has little to offer once you look away from the dilapidated hovels, dusty roads and peeling walls, sweltering humidity, and, naturally, malaria. An African town with coconut palms and a Catholic church. Even so I am drawn towards the town by the undefined tension one feels in the midriff when Clint Eastwood rides into Sergio Leone's western accompanied by a monotonous Jew's harp. I can't help turning back, time after time, to try to understand to where development has brought this town and its people.

There was a time when all the people who lived in Bagamoyo, fished, hunted and plucked the fruit from the trees, talked to each other and developed tools just as they do today. They were well adapted and qualified to overcome most threats. Even so, a small group wanted to move on and began the long odyssey which brought them north and out of Africa. Others had preceded them but did not, ultimately, survive. This group were different; perhaps they were luckier than their predecessors; perhaps they had some special characteristics. The rest of the family remained behind in Bagamoyo, where they developed what was needed and apparently led a peaceful existence for many thousands of years. The 'peaceful' is guesswork and wishful thinking, but nothing in the history, language or myths which have been preserved indicates things were in any way otherwise. The myths concern baobab trees and the human being's careless or disrespectful interaction with nature's gifts. As a rule the myths describe the people as a unit. Humanity. The human being.

Fifty thousand years later the emigrants returned to Bagamoyo as civilised, highly developed Christian and Islamic peoples with a fantastic technology and organised the greatest genocide in history. What had happened in the meantime? Which of God's gifts had brought the emigrants prosperity?

Today we regard slavery as history and of the past; an interesting subject which constitutes the weight in more or less lurid, successful books and films, but the influence of slavery on the Africa of today and African culture can hardly be overestimated.

The culmination of slavery in the seventeenth to eighteenth centuries is history's greatest holocaust, where the basis of life in whole regions of the world was utterly reduced to ruins. Gigantic areas were cleared of people, just as a rainforest is stripped of trees. Millions were taken as slaves, and still more were killed in battles with the slave-hunters. Livingstone witnessed and described in detail how slave-hunters opened fire with cannon on a peaceful market in Central Africa and slaughtered men, women and children in an inferno in which it was possible, easily and inconspicuously, to capture the surviving and unharmed.

Africa's remaining enclaves were transformed into corrupted regimes which supported the slave trade. Psychopaths and despots seized power with the slave-hunters' assistance and perverted the culture. Traditional rites were turned on their heads so that they now suddenly supported and motivated slavery. Quite paradoxically this unpleasant system became a substantial legitimising of slavery, because the wretched victims were transferred from a cabinet of horrors ruled by wild cannibals to a pious Christian existence in bondage.

As the Bishop of Copenhagen expressed it, slavery was not quite compatible with the Bible, but one ought to bear in mind that slavery removed people from an even worse existence among savage heathens in darkest Africa. Africa was destroyed. The worth of the slave trade and its additional workforce has had a monumental significance for the development of the West. The question is, whether the significant advance of the West in the eighteenth and nineteenth centuries was based, more or less, on the value of slavery. If so, then we owe an inconceivable debt to Africa. Not only did we steal everything of value, but we also removed the foundations of independent development.

Ring-tailed lemurs of Madagascar.

That one can call double theft!

Throughout the entire history of slavery there were people in the so-called developed world who opposed the institution. The Quakers in America were the first to forbid it. As far back as 1727 the trade was forbidden, and the total prohibition of the keeping of slaves came in 1761. Court cases, judgements and acts of parliament pointed for more than a hundred years in the direction of the cessation of slavery. Educated people knew perfectly well that nothing could justify the institution. The Christian ethic, which constituted and constitutes the foundation of Europe, has no place for slavery. The point is that we have not changed our minds significantly throughout the last three or four hundred years. The ethic remains the same, but where does it come from and why does it not inform all situations?

When the first modern 'double-thinking' people came to Bagamoyo they lived in small clans with a common gene pool. They exchanged genes with other scattered groups in the same manner as apes and other animals. At a certain point in time, they combined into slightly larger units which fared better than the smaller ones. Life together with other 'gene pools' has hardly been completely free of problems, since one does not have any immediate interest in supporting other gene families, and besides, one has to protect oneself against life-threatening rivals. Exactly as is the case with most other species. The value of being together, in fellowship, the common coordination of fundamental tasks of existence, has, however, overshadowed the genetic reservations. The so-called fitness was reinforced through cooperation with several of the common species. The large family has, in consequence, been more successful than the small independent groups, which, in time, disappeared.

All in all, just like a modern textbook in evolutionary biology, which maintains almost unequivocally that larger groups are more successful than small ones. Large prides of lions fare significantly better than small ones and solitaries. Hunting is far more successful, more cubs survive and, generally, the reproduction rate is far, far higher. This rule applies to all animals in social groups. As previously mentioned, the cooperation between the savannah's grazing species is another example of the power of social factors. Large groupings with partially different food niches reduce the pressure on resources and increase survival.

Our kindred forefathers had to develop a social system which could ensure coexistence and, completely in harmony with the teaching of evolution, the most social individuals stood out, because, quite simply, they fared best. The brain had already turned them into sublime solvers of problems. They could learn, remember and respond to changes, draw up new strategies, survive extremes and develop new tools. Slowly – but surely. The brain had more to offer. It could transform their people into social creatures who gradually became self-aware and conscious that fellowship brought advantages with it. The basis of what we call reciprocal altruism emerged with us. One sacrificed oneself for each other trusting that the others would do the same. If you scratch my back for five minutes, I'll scratch yours for the same length of time.

At first sight it appears to be a mutual exchange on a sensitive pair of scales, but it was regulated by social characteristics which we call feelings – characteristics which, in the final analysis, are the currency in the 'housekeeping budget'. Feelings translate actions into a human currency so there is some order in the accounts. Balance. The tools are self-awareness, cooperation and language, which unite and improve both our material and our spiritual possibilities. We think, we are and we think about reasons. The clan develops the narrative about itself in the universe, and thereby religion was on its way to establishing itself in the minds of our ancestors even though there is no complete consensus as to when this occurred.

The narrative became our common rule of conduct and assembly point and thus also the origin of what we call morals or ethics today. The narrative also became human beings' first literature. Since we could not write, but were compelled to remember, the narrative became more and more coherent and easier to remember. At the same time the brain was developed and gained an intellectual dimension. People with the best brains, which could best understand the narrative and its value, were one step ahead of all those who couldn't. Slowly but surely evolution's machine with its sureness of touch propelled the human being in the direction of greater and greater mental abilities, which could also find many other uses. We became better equipped to counteract the caprices of nature, to rearrange it and establish complex survival strategies

Spotted hyenas, young and old, emerge from their underground burrow.

such as the keeping of cattle, agriculture and the storage of various types of food.

These capabilities were developed in Africa, and the development continued among the emigrating peoples. This means that the crucial development began a long time before *Homo sapiens* migrated. This process was supported in particular by cultural activities such as cave paintings in and outside Africa, advanced burial rituals as well as high-tech tools. Advanced strategies for the obtaining of food had provided the human being with the necessary surplus capacity to cultivate the cultural and social aspect, unlike many of nature's other species which had to devote all of their waking hours to foraging.

When the brain provides its owner with maximum assistance, there arise periods of prosperity with surplus to express gratitude to 'the gods', and this is where visual art comes in. Surplus also teaches people how to create further surplus and how to store this. Cooperation across the family boundaries has created our success and hence the term 'the social human being' – far more inclusive for our species than the 'thinking human being'. This does not mean, however, that we have set ourselves apart from biology, but that the necessary altruistic, social, ethical and empathic features and skills are passed down in the biology. This mode of thinking is not especially popular because it appears to challenge the notion of free will, which is far more popular. But free will lends itself superbly well to being thought of as part of the genetic inheritance. The freedom to think, create and develop in a socially determined frame which is under the constant pressure of primeval instincts. The dualism between the individual and society, between Heaven and Earth, which seems to suffuse all human thought.

Many years after the migration from Africa, Christianity arose with the concept of 'sinful man'. We are born evil and sinful and have only limited opportunity to liberate ourselves from these far from attractive qualities. This interpretation of the human being's nature probably dates right back to the time when we consciously organised ourselves into cooperative groups. On the one hand we were Darwinian apes, where the strongest survived, mated with the most females and sent the most genes onward. On the other hand evolution also supplied us with a brain, which informed us that we could fare slightly better if we joined together in groups which worked together in hunting and protecting the group's life and territory. On the one hand we would gladly kill our neighbour and steal his food and his women, and on the other we would also use him to defend us against external enemies. Dualism, society, culture, religion and philosophy took form in the human brain.

Freud believed that the human being's instinctive behaviour was always in conflict with the advantages of organising him or herself in society. He was extremely preoccupied with this situation: one of the books in which it is described has the appropriate title of *The Burden of Culture*. He believed that the human being's primeval instincts were repressed by society, and that we became frustrated, filled with conflict and neurotic because we could not act out our primeval instincts. Culture, and thus religion have, as their purpose to keep the primeval instinct under control. Freud's theory is plausible and contains a great deal of logic, but the question is, whether it is not the exception which proves the rule: that the people he talks about are a dwindling minority of the unsuited. Primeval humans have had a pretty long time to adapt themselves to a culture of cooperation and the religious and ethical concepts have, at the same time, taken root in the human being's consciousness and in the genes. Thus it was that in the dawn of time we worked out that actions which are not in the interests of the community are evil; actions which may consider the interests of an individual here and now, but would, on the other hand, benefit neither the whole group, nor thereby the individual in the long term.

This brings us to the classic biological dilemma: if, on one side, a person behaves altruistically – self-sacrificingly – for the whole group, these qualities are wiped out in the course of time. If one does the opposite – behaves selfishly

Sable antelope suckling her calf.

– one does not advance the community, which thus fragments.

For these reasons the social society is underpinned by norms called culture, religion and ethics, which are developed and established in the genome parallel with the material development. We retain the inner survival power or primeval power in an undiluted form, but manage at the same time to channel much of it over into the community. It follows that certain ethical concepts must be universal. We possess a number of inborn behavioural norms, which, in consequence, are passed down in human nature – i.e. genetically in one or another sense – which corresponds to the norms which are described in, among other places, the Ten Commandments. We no longer need to learn that thou shalt not kill, steal or envy thy neighbour. Certain of these norms exist independently of the individual person or society and do not permit themselves to be 'relativised'.

When, anyhow, we must conceive of ourselves as sinful or evil, it is precisely because we know well what is right and ethical – that which lies deep within us – but on the other side we also have to live with the urge to act out primeval instincts, to advance our own genes at the expense of the community. This is

not, however, something which will make us ill as Freud maintains.

The myth which describes the human being's dark compulsions we call 'the Fall'. Protestantism and Catholicism each tackle the consequences of this in their particular ways. The Catholics neutralise the human being at baptism, and one becomes oneself master of one's own actions, whereas the Protestants acknowledge that the indwelling forces are too strong for an individual to be given the entire responsibility. Salvation can only be achieved through faith. Thus it is that the simple survival instincts which brought us down from the trees to fight side by side and form societies were reflected in complex religious conceptions. At first glance, the Catholic interpretation may seem to be more sympathetic because the human being must assume a personal responsibility, which, however, is not so great that one cannot, through confession and appropriate contrition, obtain absolution – Lutherans are supposed only to believe, but there's the crux: if one truly believes, this also involves accepting the Christian ethic without question or doubt. Therefore the possibilities for an ethical life are far greater for Protestants because if a Protestant betrays his faith, he betrays his ethics and his basis of life at the same time – while a Catholic, on the other hand, can be content with being a poor sinner with infinite possibilities of forgiveness.

Just as the human being is a result of evolution, religion, ideas and culture

are results of the same. Ideas reflect the actual productive conditions, the community, precisely as Marx and his faithful followers believed.

Through history, the human being has developed ideas about cooperation and coexistence which are deeply anchored in the genome, and which can be traced back to the earliest civilisations. Therefore we also want to be able to demonstrate the presence of general ethical traits in every human being. The people with whom this has not been the case died out along the way. Gradually, as the human being has spread over vast areas under extremely varied conditions of life, the religious and ethical traits have adapted to actual situations, but, however, no more than that we can rediscover common features.

Marx characterises every society against a background of its form of production: slave society, feudal society and capitalism. For this so-called basis there is a corresponding superstructure which contains the prevailing ideas, culture, religion and ethics. In the case of Marx it was important to point out that this superstructure served to conceal the truth – to veil reality – from the people and socialise them into accepting a fundamentally unjust society. Curiously enough this problem was supposed to disappear with the advent of the socialist or communist society, where the culture now suddenly became something which was in the best interests of the people. Even though the theory limps painfully along in this context, it does reflect the connection between practice and ideas. At first glance it appears extremely reasonable, even though one should probably distinguish between topical ideas which underpin a changeable society, and the supporting pillars which hold up the whole of humanity. The universal, wherever that's supposed to come from.

That which we call development seems to follow the same rules as evolution, and they are two sides of the same coin. How and how fast human development leaves its impression on the genetics is something about which we still know very little, but, as I hinted earlier, it may move somewhat faster than we imagine. Accordingly, both development and evolution seem to work with a change of tempo against a background of given preconditions, even though there will naturally be examples of changes in environment happening so fast that evolution cannot succeed in compensating. A concrete example is the drastic fall in diversity which follows the removal by civilisation of habitats.

Consequently our development appears to follow a consistent track, where technological gains are followed by cultural adaptation which leaves its mark in the genome.

Neither human beings nor animals develop directionally. It's not on the cards that a rice-growing culture in Southeast Asia, a nomad culture in Africa or a hunting culture in the polar regions are on the same route towards the same development as us. Today many people assume – just like Marx – that we are on our way in a specific direction, and that societies pass through a natural development towards industrial society, but it is the modern world which influences the other cultures and not they themselves which pass through a desired development. Animals and human beings have a tendency to optimise the current living conditions. For some species this means development at a rate of knots and for others (and peoples) a long, conservative development. Perhaps all development happens in a multitude of tempi. The result is long periods with something that approximates stasis and others with tiger leaps.

When things go well for a sufficiently long period of time, routines are worked where the status quo is maintained; it goes well when we do what we're used to. Conservatism becomes the equivalent of security. The cattle nomads in East Africa have learned to live with a particular cycle of drought and the strategy for handling the drought is quite clear. Right up to the present day there are examples of those staunchly adhering to tradition doing better than those who give themselves over to the help and technology of modern society. Along the way some will die depending on the character of the drought. All this must be taken into consideration in their culture. Myths and rituals deal with the issues which are infused in new generations with their mothers' milk. One survives by listening to experience, therefore it is a society where the eldest have power and respect. The society is conservative, but its cohesive force is enormous. It is, so to speak, Africa's answer to the fishermen of Denmark's

Indre Mission (a Christian sect of very austere, fundamentalist beliefs) which Danish author Hans Kirk has described in his novels from the West Coast of Jutland.

When we returned to Bagamoyo to slaughter our ancestors, despite thousands of years of learning and social development, we may have developed into other people. The great world had brought about many changes. The innovative problem-solvers who abandoned Africa had developed in a more aggressive and less empathic direction towards the members of other clans. The social element within the clan had certainly expanded, but, on the other hand the restrictions determining who was inside and who was out had been clearly laid down. Great civilisations had already developed and had collapsed again. Through the art of cooperation we built Babylon and rich cultures in the lush landscapes between the Euphrates and the Tigris which today are no more than barren desert on a par with the other high cultures of the Middle East and Northeast Africa. We did not understand the limitations of nature and we harvested and harvested until nothing was left. We left behind us treeless, infertile and exhausted soil. Next we turned our attention to our neighbour's fields. And that paid off!

Among Africa's dominant, original peoples, the Bantu, there is no actual warrior nation, apart from a few South African tribes who formed the Zulu Kingdom under pressure from the slave traders. Everything suggests that Africans experienced a peaceful existence in close contact with nature until they were confronted with the returning emigrants who were either white Europeans, Arabs or nomadic tribes, which had also probably been among the immigrants. Is it possible that there is a crucial difference between the original Africans and all the immigrants, and in which case is such a difference genetically or culturally determined?

We know that all peoples outside Africa are more closely related to each other than they are to the Africans, that small genetic differences can have an enormous significance, and that these can propagate rapidly in a population. We have seen how a Siberian fox can change personality in the course of a few generations with artificial selection. We know that there are significant differences in intelligence between the populations of the Earth, and that the human being, seen in general, is far more aggressive than its close relative the bonobo. We also know that our history outside Africa is one long war of aggression, so there is much that suggests that the social civilisation gene has acquired a consolation prize of which we can undoubtedly be less proud.

Our exploration and investigation of our part of the planet has developed us in various directions. We have changed external characteristics such as skin, hair and colour of eyes, growth of beard, musculature, fat distribution and bone structure. In addition there are a large number of other physical differences, such as milk tolerance, the ability to break down alcohol and a mass of other more or less fortunate genetic variations which offer a picture of the human being's colonisation of the Earth. Apparently we know a lot about how the so-called intelligence factor is distributed among the Earth's inhabitants, but here, unfortunately, is where the party has to stop, and our knowledge fades from sight like the mountain gorillas in the mist. There are, however, no grounds to believe that many of the characteristics for which we are looking in – for instance, personality tests – are evenly distributed among human beings. The question is, which characteristics have had the greatest success? And which will produce the greatest success in the future? Is the Islamic Sultanate better at ensuring human survival than democracy? Does aggressive force create better fitness than empathic, social cooperation?

The formulation of, and the adoption in law, of human rights contains a justified hope that the social human being will win the battle, even though it is the very few who understand that these fundamental rights can only be realised if we can agree to assume, collectively, the full responsibility for Bagamoyo.

Redbilled hornbill.

Sources

As is apparent from the foreword and the literary references that I have, of course, used many sources: textbooks, reference works and scientific articles have contributed with numerous facts and thoughts. This book is, however, not science and therefore there is no reason to reduce the readability with source acknowledgement.

In connection with the final proof-reading I googled almost all the factual information and confirmed that the scientific community has penetrated so deeply inside us that one cannot actually conceal anything at all. At the speed of light a few searches turn up the background for anything at all. Wikipedia, Google Scholar and the Encyclopedia contain answers to everything about which I've written. Wow, it's as simple as that! Try searching on Vitamin D. Daunting reading awaits you, and before you've looked around you fill yourself with 50 micrograms daily through all nine months of the Danish winter. But this may also be one of the best things you can do for your health.

Sources do not solve all your problems however. I have on many occasions pointed out disagreement over numbers and dates, and that will be rapidly confirmed if anyone poaches on my preserves. When did the human being migrate out of Africa? How did we develop into cultural personalities? How much do we resemble apes? In what sequence did the whales develop? Many pieces of potsherds, genetic dating, carbon 14 and structural traits produce a total picture, but unfortunately the various bits of information do not always pull in exactly the same direction. My method is to use the most meaningful: that which results in a reasonable picture, that which cannot be said to be wrong.

How much do we resemble apes? The genetic adding machine is pretty imprecise. Partly we need to know more about the consequences of the actual differences, and partly they are only observations of averages, based on a selected part of the genome. If one compares the genomes from a South African Bushman with those of an inhabitant of Easter Island, one will probably find enormous differences. Half a per cent, perhaps? What number shall we compare with the apes? It is, in any case, not especially meaningful to measure hundredths of a per cent, unless one understands the context. No scientific number can be read without methodical explanations. Might there not be a significant risk factor with 'Google science'? Anyhow, enjoy yourselves.

Acknowledgements

Dr.scient Kay Petersen; Mag.art. Torben Holm-Rasmussen; Former Danish High Court Judge Mogens Hornslet must be thanked for non-committal reading, helpful comments, suggestions and corrections. Marketing Director Jesper Dichmann for prodigious enthusiasm and praise (and one certainly can't have too much of that). My publishing editor Ole Jørgensen deserves my sincerest gratitude, additionally for the palpable enthusiasm behind his top professional reading. My girls in the natural science department, Berit Willumsgaard (my wife) and Cæcilie Willumsgaard (my daughter) are to be thanked for reading, listening, discussing, searching for sources and providing peace and quiet when the genius was at work and a whole lot more besides.